WIRING HANDBOOK FOR RURAL FACILITIES

WIRING HANDBOOK
for Rural Facilities

TECHNICAL EDITOR
LaVerne Stetson

MWPS
MidWest Plan Service
A Foundation of Knowledge

MWPS-28

Copyright© 2013, Iowa State University / MidWest Plan Service
All rights reserved. Fourth edition.

Reproduction of this publication or any part therein is prohibited without the written approval of MWPS. Requests should be sent to MWPS. In your request, please state which parts you would like to reprint and describe how you intend to use the material.

Designed by Kathy J. Walker

Publisher

MidWest Plan Service (MWPS)
122 Davidson Hall
Iowa State University
Ames, Iowa 50011-3080
Phone: 515-294-4337 / 800-562-3618
Fax: 515-294-9589
E-mail: mwps@iastate.edu
Web site: www.mwps.org

Fourth edition, 2013

Library of Congress Cataloging-in-Publication Data
Wiring handbook for rural facilities / technical editor, LaVerne Stetson.—4th ed.
   p. cm.
   "MWPS-28."
   Includes bibliographical references and index.
   ISBN 0-89373-108-0 (alk. paper)
1. Electricity in agriculture--Handbooks, manuals, etc. 2. Electric wiring, Interior—Handbooks, manuals, etc. I. Stetson, LaVerne E. (LaVerne Ellis) II. Midwest Plan Service.
   TK4018.F287 2013
   621.319'24—dc23

2012049082

The last number is the print number
10 9 8 7 6 5 4 3 2

Technical Editor

LaVerne E. Stetson, P.E., Adjunct Professor Emeritus,
   Biological Systems Engineering, University of Nebraska-Lincoln, and
   Agricultural Engineer (retired), USDA-ARS

Contributions to this Edition

Dr. Robert J. Gustafson, Director EEIC, The Ohio State University of Ohio
Dr. Douglas J. Reinemann, Professor, Biological Systems Engineering,
   University of Wisconsin

Contributions from the Third Edition

This publication updates the Third Edition of the Wiring Handbook for Rural Facilities by incorporating changes to the National Electrical Code (NEC) published in 2011. This fourth edition builds on an original text developed by MPWS electrical Systems Work Committees and on the previous technical and editorial contributions of the following:

Bill Koenig, P.E., Former staff engineer and project manager, MWPS, Iowa
   State University.
Dr. Carl Bern, P.E., Professor, Agricultural and Biosystems Engineering,
   Iowa State University
Dr. Tom Bon, P.E., Professor, Agricultural and Biosystems Engineering,
   North Dakota State University
Dr. Ted Funk, P.E., Extension Specialist, Agricultural Engineering,
   University of Illinois
Dr. Doug Reinemann, Professor, Biological Systems Engineering,
   University of Wisconsin-Madison
Dr. Joe Zulovich, P.E., Structures Engineer, Division of Food Systems and
   Bioengineering, University of Missouri-Columbia
Letitia Wetterauer, Technical Illustrator, Alpine, Texas

Additional Support for this Edition

Jack Moore, Editor

. . . And Justice for All.
Iowa State University / MidWest Plan Service does not discriminate on the basis of race, color, age, religion, national origin, sexual orientation, gender identity, sex, marital status, disability, or status as a U.S. veteran. Inquiries can be directed to the Director of Equal Opportunity and Diversity, 3680 Beardshear Hall, 515-294-7612.

CONTENTS

Acknowledgments		vii
List of Figures		viii
List of Tables		x
Chapter 1	Electrical Design and Component Basics	1
	1	
	3	
	3	
	s, 4	
	s, 11	
	ıt Guidelines, 15	
		19
	19	
	19	
	2	
	otection, 27	
	Means, 28	
	ction, 31	
	ductors, 32	
		35
	35	
	Breakers, 35	
	it Breakers, 35	
	ircuit Interrupters, 35	
	nel (SEP), 36	
	iring Diagram, 40	
	ce Entrance, 40	
	trode Conductor, 41	
	trode System, 42	
	, 43	
	44	

TITLE:
Dairy Freestall Housing and Equipment
MWPS-7
Eighth Edition
ISBN 0-89373-109-9

NEW:
Major revisions and expansions in seven of its ten chapters, this edition has more than doubled in size and information!

SIZE:
232 pages, 8.5 x 11, softcover, illustrated, tables, references, index

COST: $35.00

ARRIVAL: May, 2013

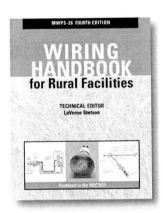

TITLE:
Wiring Handbook for Rural Facilities:
MWPS-28
Fourth Edition
ISBN 0-89373-108-0

NEW:
Updated for the 2011 National Electrical Code (NEC). Includes lighting options and enhanced stray voltage information.

SIZE:
96 pages, 8.5 x 11, softcover, illustrated, tables, references, index

COST: $22.00

ARRIVAL: February, 2013

Chapter 4. **Standby Power** 47
  Generator Types, 47
  Generator Sizing, 48
  Installation, 50
  Operation, 52
  Maintenance, 52

Chapter 5. **Alarm Systems** 53
  Homemade Alarms, 53
    Battery-Operated, Relay-Controlled Alarm, 53
    Solenoid Valve-Controlled, Compressed Gas Horn, 53
    Alarm for Multiple Fans, 53
    Battery-Operated Alarm with Thermostat, 53
    Combination Alarm System, 55
    Battery Maintenance, 55
  Fire Detectors 55

Chapter 6. **Stray Voltage** 57
  Causes, 58
    On-Farm Sources of Stray Voltages, 59
    Off-Farm Sources of Stray Voltages, 62
  Solutions, 64
    Eliminate or Reduce the Stray Voltage, 64
    Isolate Voltage, 64
    Install Equipotential Planes, 65

Chapter 7. **Lightning Protection** 71
  Air Terminals, 72
  Main Conductors, 72
  Secondary Conductors, 73
  Arresters, 74
  Ground Connections, 75
  Metal-Clad and Steel-Framed Buildings, 76
  Fences, 77
  Trees, 77

Chapter 8. **Example Buildings** 79
  Example: 8-1. Machine Storage Area, 80
  Example: 8-2. Livestock Building, 84

  Glossary 89
  Additional References 90
  Index 91

ACKNOWLEDGMENTS

MWPS would like to thank the following group for their major financial support towards the revision of the Third Edition and this Fourth Edition:

Midwest Rural Energy Council (MREC)
www.mrec.org

MWPS would like to thank the following groups for their financial support of this project:

Alliant Energy
www.alliantenergy.com

Focus on Energy
www.focusonenergy.com

Minnesota Power
www.mnpower.com

Otter Tail Power Company
www.otpco.com

Touchstone Energy
Cooperatives of Iowa
www.touchstoneenergy.coop

We Energies
www.we-energies.com

Wisconsin Public Service (WPS)
www.wisconsinpublicservice.com

Xcel Energy
www.xcelenergy.com

MWPS would like to thank the following businesses and people for providing the equipment or photos shown in the book:

Stephen Brunia and Bradley Brunia
Electric Wholesale Co.
Ames, Iowa
www.ewci.net

Lowe's of Ames
Ames, Iowa

Carl Bern, Professor
Iowa State University

Ted Funk, Extension Specialist
University of Illinois

LaVerne E. Stetson, Agricultural Engineer (retired)
University of Nebraska-Lincoln and USDA-ARS

LIST OF FIGURES

Chapter 1. **Electrical Design and Component Basics**
- **1-1.** Required overhead line clearances on loading and non-loading sides of grain storage bins or buildings. 2
- **1-2.** NM and UF cable. 3
- **1-3.** Example of a surface mount plastic fixture that should NOT BE used in damp or dusty environments. 3
- **1-4.** Incandescent fixture for damp buildings. 4
- **1-5.** Fluorescent fixture for damp buildings. 4
- **1-6.** Outlet boxes, junction boxes, and access fittings. 5
- **1-7.** Dustproof and watertight boxes with appropriate switch covers or receptacle covers. 5
- **1-8.** Watertight cable-to-box connection. 6
- **1-9.** Examples of equivalent number of conductors in boxes. 7
- **1-10.** PVC couplings and threaded adaptors. 9
- **1-11.** Fasteners for PVC conduit. 9
- **1-12.** Premade PVC conduit bends. 9
- **1-13.** Allow for thermal expansion in PVC conduit. 10
- **1-14.** Typical PVC conduit-to-box connection. 10
- **1-15.** Liquid-tight flexible nonmetallic conduit. 11
- **1-16.** Electrical equipment symbols. 11
- **1-17.** Incandescent bulbs. 12
- **1-18.** Fluorescent tube. 14
- **1-19.** Compact fluorescent bulbs. 14
- **1-20.** High intensity discharge fixture. 14
- **1-21.** Fluorescent and compact fluorescent lights used in a farrowing room. 15

Chapter 2. **Branch Circuits**
- **2-1.** Light with one switch. 20
- **2-2.** Circuit wiring schematics. 20
- **2-3.** Maximum number of lights or duplex convenience outlets per general-purpose circuit. 21
- **2-4.** Typical receptacles and plugs for agricultural-type buildings. 21
- **2-5.** A circuit for portable motors one-third horsepower or under. 22
- **2-6.** Motor circuit. 23
- **2-7.** Ventilating fan motor circuit. 24
- **2-8.** Motor circuit. 25
- **2-9.** Ventilating fan motor circuit. 26
- **2-10.** Timer wired in parallel with thermostat. 27
- **2-11.** Typical short-circuit protection for smaller motors. 28
- **2-12.** AC snap switch. 31
- **2-13.** Knife switch with fuse holders. 31
- **2-14.** Motor starters. 31
- **2-15.** AWG wire sizes. 33

Chapter 3. **Service Entrance**
- **3-1.** Wiring of branch circuit breakers in SEP. 36
- **3-2.** GFCI circuit breaker. 36
- **3-3.** All cable and conduit enter SEP at or near the bottom. 36
- **3-4.** Service entrance panel (SEP) box. 37
- **3-5.** Wiring sub panels. 38
- **3-6.** Electrical grounding system. 40
- **3-7.** Grounding the service entrance panel to the neutral/grounding bar with a bonding strap. 41
- **3-8.** Ground rod installation. 43
- **3-9.** Power route from plant to load. 44

LIST OF FIGURES

Chapter 4.   **Standby Power**
- **4-1.** Generator types. 47
- **4-2.** Double-pole, double-throw transfer switch. 51

Chapter 5.   **Alarm Systems**
- **5-1.** Battery-operated, relay-controlled alarm. 53
- **5-2.** Solenoid valve-controlled, compressed gas horn. 54
- **5-3.** Alarm on multiple-fan system. 54
- **5-4.** Battery-operated alarm with thermostat. 55
- **5-5.** Temperature- and power-sensitive alarm. 55

Chapter 6.   **Stray Voltage**
- **6-1.** Example of stray voltage. 57
- **6-2.** Neutral and equipment grounding system. 58
- **6-3.** Ground-fault. 59
- **6-4.** Equipment grounding reduces stray voltages. 60
- **6-5.** Ground-fault from an underground cable. 61
- **6-6.** Improper connection of a circuit neutral and equipment ground. 61
- **6-7.** Unbalanced 120-volt loads create current on the secondary neutral. 61
- **6-8.** Balanced 120-volt loads at the service entrance panel. 62
- **6-9.** Ground-fault at a neighboring farm. 63
- **6-10.** Primary neutral current returning to substation via a nearby farm. 63
- **6-11.** Distribution transformer with isolating device between the neutrals. 65
- **6-12.** Transformers for electrical isolation. 66
- **6-13.** Grounding non-electrified metallic equipment. 65
- **6-14.** Grounding waterers. 67
- **6-15.** Equipotential plane in stanchion and tie-stall barns. 68
- **6-16.** Equipotential plane in milking parlor. 69
- **6-17.** Equipotential planes in swine buildings. 70

Chapter 7.   **Lightning Protection**
- **7-1.** A typical air terminal. 72
- **7-2.** Common conductors. 72
- **7-3.** Conductor bends. 73
- **7-4.** Lightning protection for barns. 73
- **7-5.** Lightning protection for houses. 74
- **7-6.** Lightning arrester. 74
- **7-7.** Ground rods. 75
- **7-8.** Grounded stranded cables. 75
- **7-9.** Counterpoise ground. 76
- **7-10.** Steel frame as conductors. 76
- **7-11.** Grounding wire fences. 77
- **7-12.** Grounding trees. 77

Chapter 8.   **Example Buildings**
- **8-1.** Example machine storage shop building electrical system layout. 80
- **8-2.** Electrical plan for example machine storage building. 81
- **8-3.** SEP wiring schematic for example machine storage. 84
- **8-4.** Example swine farrowing building layout. 85
- **8-5.** Electrical plan for example farrowing building. 85
- **8-6.** SEP wiring schematic for example swine building. 88

LIST OF TABLES

Chapter 1. **Electrical Design and Component Basics**
    1-1. Minimum clearance distances for grain storage bins or buildings. 2
    1-2. Box volume required for conductors. 6
    1-3. Metal box volume and equivalent conductor capacity. 8
    1-4. Conductor insulation classifications. 9
    1-5. Maximum number of conductors in Schedule 40 rigid PVC conduit. 9
    1-6. Expansion characteristics of PVC rigid nonmetallic conduit. 10
    1-7. Characteristics of lights. 13
    1-8. Interior illumination levels and light energy requirements for agricultural-type buildings. 16-17
    1-9. Interior light energy of high-intensity discharge lights for agricultural-type buildings. 17
    1-10. Typical illumination levels. 17
    1-11. Guidelines for locating lights in agricultural-type buildings. 17-18
    1-12. Guidelines for locating duplex convenience outlets (DCO) and special purpose outlets (SPO) in agricultural-type buildings. 18

Chapter 2. **Branch Circuits**
    2-1. Typical equipment for motor circuit devices. 26
    2-2. Maximum rating of motor short-circuit and ground-fault protection device. 27
    2-3. Full-load currents for single-phase AC motors (amps). 28
    2-4. Assigned horsepower ratings for attachment plugs and receptacles. 30
    2-5. Effect of voltage drop on power and light loss. 33
    2-6. Minimum UF and NM cable copper conductor size for branch circuits. 33
    2-7. Minimum THW, THWN, RHW and XHHW copper conductor size for branch circuits. 34

Chapter 3. **Service Entrance**
    3-1. Ampacity of copper conductors (amps). 35
    3-2. Demand factor. 39
    3-3. Sizing grounding electrode conductors. 42
    3-4. Amperage capacity of service entrance conductors. 44
    3-5. Aluminum wire size for a 2% voltage drop. 45

Chapter 4. **Standby Power**
    4-1. Typical equipment wattages. 49
    4-2. Motor wattages. 49

CHAPTER 1

Electrical Design and Component Basics

THIS HANDBOOK OUTLINES materials and methods for electrical equipment and wiring in rural buildings. It also can help determine if existing wiring is adequate. It does **not** cover wiring from the power supplier to the watt-hour meter. The rural electrical distribution and service entrance equipment must be installed by a qualified electrician cooperating with the power supplier.

Before beginning any project, check with local rules to determine if you can do your own wiring. There may be conditions where this is prohibited. Voltages from rural electrical systems can cause electrocution, so be sure you understand how to work safely with electricity. High voltage conductors leading to a transformer are bare and especially dangerous. Follow the **10-foot rule** when working around overhead power lines. The 10-foot rule is not a code-defined rule, but it is a good safety guideline when operating near high-voltage conductors.

You will also need to consult with the following persons and agencies to ensure proper wiring:
- The power supplier to help plan the distribution system to the building.
- A qualified electrician to help plan and install service entrance panels and motor circuits, to select conductors and fixtures, and to verify compliance with state and local codes.
- An electrical equipment supplier for dustproof and watertight fixtures, and for wiring required for damp locations. This equipment may be available only through electric wholesale supply stores. Plan ahead; you may need to order equipment.
- Your insurance company to help meet insurance requirements. If the electrical system does not meet the company's standards, the insurance company may increase your rates or refuse to insure the building.

Refer to the **Glossary** for definitions of unfamiliar terms.

Codes and Safety

The code referred to for electrical work in the United States is the NFPA 70-2011 *National Electrical Code (NEC)*, registered as a trademark and published by the National Fire Protection Association (NFPA), Quincy, MA 02169. The *NEC* is a guide to proper and safe materials and installation methods. Even though many rural buildings do not presently fall under *NEC* jurisdiction, it is imperative to follow the *NEC* for minimum safety reasons. Also, your insurance company may require an installation that meets *NEC* standards for approval. Before starting construction, check to see if a wiring permit is required.

This handbook closely follows the 2011 edition of the *NEC* for the parts that apply to typical rural buildings. Occasionally, the *NEC* may have more exceptions to the regulations than stated in this handbook. In some cases, this handbook may be more stringent than the *NEC*.

The text occasionally references an *NEC* code (for example, *NEC* 547.9) where 547 is the *NEC* Article and 547.9 is an *NEC* section.

Each year there are electrical fatalities from contact with overhead power lines. Loading or unloading of grain bins is particularly hazardous when overhead power lines are present. The Institute of Electrical and Electronic Engineers (IEEE) has developed a code called the National Electrical

Safety Code (NESC), part of which is used by electric utility companies to locate power lines. The NESC has rules to keep power lines a minimum distance away from grain storage structures. Those distances are listed in Table 1-1 and illustrated in Figure 1-1. Contact your power supplier if the power lines located near your grain handling facilities or near similar facilities do not have the proper clearance, or if you need help interpreting NESC clearance requirements. Also, contact your power supplier when you are choosing a site on which to build new facilities.

Materials

Use only equipment listed by Underwriters Laboratories Inc. (UL) or an equivalent laboratory. The UL is a nonprofit organization that establishes standards and tests for various products. The UL does not approve items but rather lists them as having met minimum safety standards. Listed items include switches, plugs, connectors, receptacles, and panelboards. Equipment also must be installed according to the listing or labeling included with the product.

Use a minimum of 15-amp (ampere) rated materials for general lighting and general-purpose circuits. Use 20-amp rated materials for locations where rugged service is required. A bonding (grounding) conductor is required in all circuits whether in a cable or conduit.

Table 1-1. Minimum clearance distances for grain storage bins or buildings.
Use with Figure 1-1.

Total bin height (V) (feet)	Minimum clearance (H) (feet)
15	33
20	38
25	43
30	48
35	53
40	58
45	63
50	68
55	73
60	78
65	83
70	88
75	93
80	98

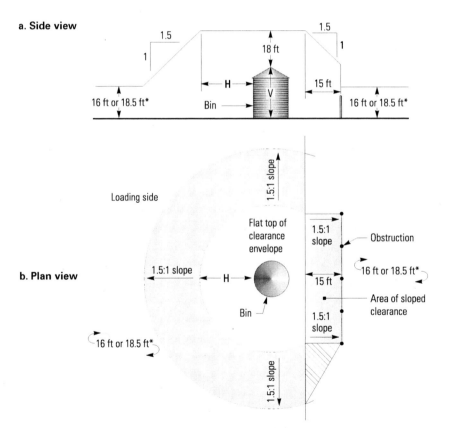

Figure 1-1. Required overhead line clearances on loading and non-loading sides of grain storage bins or buildings.
*16 feet for secondary duplex, triplex, or quadplex; 18.5 feet for phase conductors for circuits up to 22 kV.

Based on National Electrical Safety Code Article 234, Section F and Rule 232.
A non-loading side is an area in which an auger, conveyer, or other type of filling equipment cannot operate due to an obstruction, such as a permanent fence or building. Minimum clearance is for the largest final power line sag.

Building Groups

This handbook applies primarily to agricultural-type buildings not used by the public. In this book, an agricultural-type building is defined as a structure for housing farm implements, hay, grain, animals, or other agricultural produce. It is not a building used for human habitation or a place of employment where agricultural products are processed, treated, or packaged.

Agricultural-type buildings are divided into three groups:
1. **Dry:** machine storage buildings, shops, and garages not attached to the residence, *NEC Chapters 1 to 4.*
2. **Damp:** livestock housing (open or closed), milking centers, ventilated manure pits, well pits, silos, silo rooms, and high-humidity produce storages (for example, places to store potatoes and apples). It also includes buildings or areas that are washed periodically, *NEC 547.1 (B).*
3. **Dusty:** fertilizer, dry grain, and dry hay storage buildings; and grain-feed processing centers, *NEC 547.1 (A).*

Not included in this handbook are structures housing methane or alcohol production equipment or any other system that may produce explosive gas or dust. Those buildings may require explosion-proof materials; for more information refer to the *NEC Chapter 5.*

Different parts of the same building may fit different groups. For example, a wash-down area in a machine shed fits the **Damp** group while the rest of the building fits the **Dry** group.

Dry Buildings

Dry buildings (dry 100% of the time) do not require special wiring materials, but use quality materials and practices as outlined in the *NEC*. Generally, dry buildings may be wired with the same type of materials used for most residential wiring.

Surface wiring is recommended and saves materials and labor. Install the cable and components in high or out-of-the-way areas to reduce the possibility of physical and rodent damage. Install cable in conduit in areas subject to damage. Type **NM** cable is acceptable, but use only indoors and in areas that do not experience long periods of high relative humidity. Use type **UF** for damp or wet indoor locations, underground use, and outdoor locations. Figure 1-2 depicts **NM** and **UF** cables. Wiring installed outdoors must be labeled as sunlight resistant.

Agricultural-type buildings that can be wet, damp, or dusty will require dustproof and watertight fixtures that resist physical damage. Metal or plastic boxes are acceptable as long as they are dustproof and watertight. These types of boxes have screwed-on covers. Plastic boxes for use with **NM** clips are not acceptable. Do not use the less expensive type of surface mount fixtures, such as those illustrated in Figure 1-3, in new wiring installations.

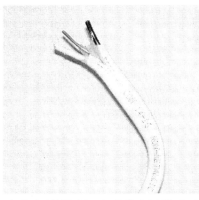

a. Conductors are nonmetallic (NM), and have a plastic, flame-resistant, and moisture-resistant sheath.

b. Conductors can be used as an underground feeder (UF) and branch circuit cable. It has a flame-resistant and moisture-fungus-corrosion resistant cover.

Figure 1-2. NM and UF cable.
Use cable with a bonding (grounding) conductor.

Figure 1-3. Example of a surface mount plastic fixture that should NOT BE used in damp or dusty environments.

Damp Buildings

Damp buildings require special materials and wiring methods because high levels of moisture and corrosive dust and gas can quickly corrode standard electrical equipment. Dust and moisture accumulation leads to fire and safety hazards by creating short-circuits or heat buildup in electrical fixtures. All wiring boxes and fixtures must be dustproof and watertight and made of corrosion-resistant materials.

Lighting

For incandescent-type lights, use dustproof and watertight fixtures with heat-resistant globes to cover the bulb, as seen in Figure 1-4. For fluorescent lights, use dustproof and water-resistant fixtures with a gasketed cover, as shown in Figure 1-5. Make sure the fixture wiring is temperature rated to allow use of **UF** cable. **UF** cable is rated for up to 194 °F.

Surface mount cable

For damp buildings, use type **UF** cable **with a bonding (grounding) conductor**, as illustrated in Figure 1-2.

Mount cable on the inside surface of walls or ceilings with plastic coated staples or plastic straps with corrosion-resistant nails at least every 4 1/2 feet and within 12 inches of junction or fixture boxes (*NEC 334.30*). Install cable where it cannot be easily damaged. Avoid sharp bends in the cable—minimum radius is 5 times the largest diameter of the cable.

If the building has joists, run the conductor along a joist or a beam at least 2 inches from the bottom. If the conductor must run perpendicular to joists or ribs of metal liners, install a 1- by 2-inch running board on the bottom of the joists or ribs, and attach the cable to the board.

Boxes

Enclose every wire splice, switch, and receptacle in a box. Mount every light fixture on a box, *NEC 300.15*. All boxes are to be noncorrosive and dustproof and watertight. Use molded plastic boxes. Gasketed covers, like those shown in Figure 1-6, are required to seal all junction boxes. Use receptacle boxes equipped with gasketed, spring-loaded covers, *NEC 547.5(C)*. Switch boxes can have spring-loaded covers, watertight switch levers or watertight covers over the surface. Figure 1-7 shows a variety of this type of box.

Mount boxes where they are protected from animals and moisture. *NEC 300.14* requires at least 6 inches of free conductor beyond the cable sheath in the box when running cable into boxes. It is important to have at least 3 inches of conductor extend outside of the box. You may also need extra conductor to replace switches, receptacles, or fixtures later. Mount switch boxes on the latch side of a doorway for easy access.

Solderless connectors (wire nuts) are one of the most common ways to splice conductors. They are available in various sizes, depending on the number and size of conductors to be joined. Use the correct connector for your application. There are special wire nuts for wet locations such as motor terminations in pump areas. Make sure solderless connectors are twisted until they hold every conductor tightly. Tug on each conductor before you leave the connector.

Use dustproof and watertight cable-to-box connectors, like those shown in Figure 1-8. They have tapered hub

Figure 1-4. Incandescent fixture for damp buildings.
Fixtures are nonmetallic, globed, and dustproof and watertight. Use fixtures with at least a 150-watt rating.

Figure 1-5. Fluorescent fixture for damp buildings.
Fixtures are nonmetallic, enclosed, and gasketed.

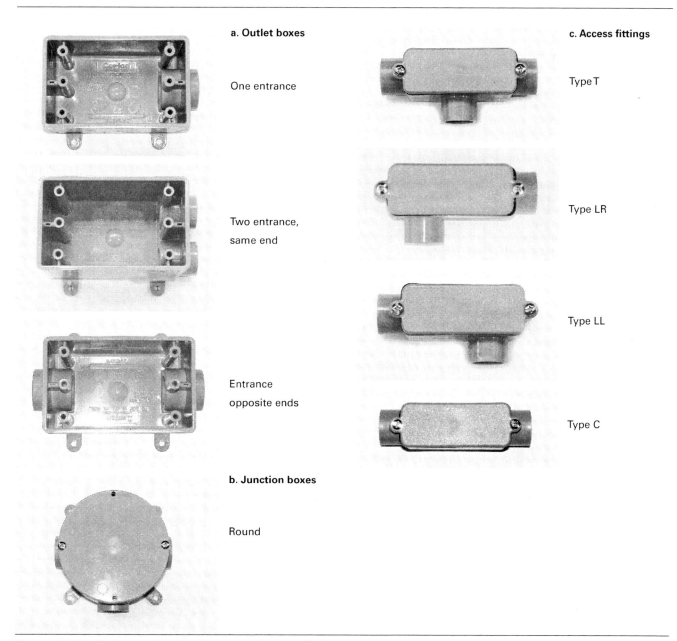

Figure 1-6. Outlet boxes, junction boxes, and access fittings.

Figure 1-7. Dustproof and watertight boxes with appropriate switch covers or receptacle covers.

CHAPTER 1. ELECTRICAL DESIGN AND COMPONENT BASICS

threads and a rubber, neoprene, or plastic bushing with an oval hole to fit the cable size used. When connected to a box, the bushing is compressed to form a watertight seal. Use boxes or adapters that are threaded for the tapered hub connectors or those that have other provisions for a liquid-tight seal. Figure 1-14 illustrates a PVC conduit-to-box connection that uses a sealing washer to create a liquid-tight seal.

A box must contain adequate volume for all the conductors and devices it will house, *NEC 314.16*. Crowding conductors into boxes makes working difficult, increases work time, concentrates heat, and makes short-circuits more likely. Based on the largest conductor entering the box, determine box size by multiplying the equivalent number of conductors by the volume/conductor values in Table 1-2. Not only are the conductors counted, but each type of device, such as a cable clamp, that takes up space in the box is counted as a conductor to allow for the space it uses, *NEC 314.16(B)*. Use the following guidelines for counting the equivalent number of conductors:

Table 1-2. Box volume required for conductors.
Based on *NEC Table 314.16 (B)*.

Conductor size (AWG)	Box volume (cubic inches per conductor)
14	2.00
12	2.25
10	2.50
8	3.00
6	5.00

- Each conductor passing through a box without being spliced or connected to a device is counted as one conductor.
- Each conductor connecting to a splice or device is counted as one conductor. However, if a conductor is completely contained within the box, such as a pigtail connection, it is not counted.
- All grounding conductors in a box are counted as only one conductor unless a second set of equipment grounding conductors (as permitted by *NEC 250.146(D)* dealing with isolated equipment grounding conductors) is present in the box, then an additional conductor shall be counted.
- A switch or receptacle is counted as two conductors.
- Each of the following types of fittings is counted as one conductor: cable clamps, fixture studs, hickeys, and straps. Each type of fitting is counted as one even if there is more than one fitting of that type. But, if there is more than one type of fitting, each type must be counted as one. For example, if a box has two cable clamps and one fixture stud, the fittings add up to an equivalent of two conductors of the size connected to that fitting.

Figure 1-9 gives the equivalent number of conductors in some example boxes. Table 1-3 gives the volumes of some common metal boxes and the maximum number of equivalent conductors. Nonmetallic boxes usually have the volume in cubic inches (in³) stamped inside the box.

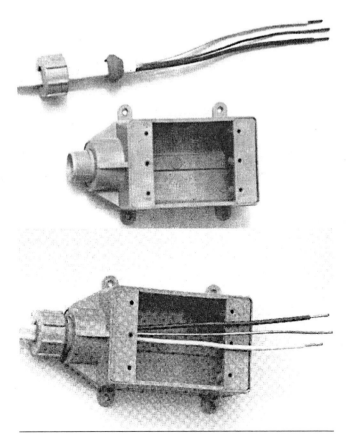

Figure 1-8. Watertight cable-to-box connection.

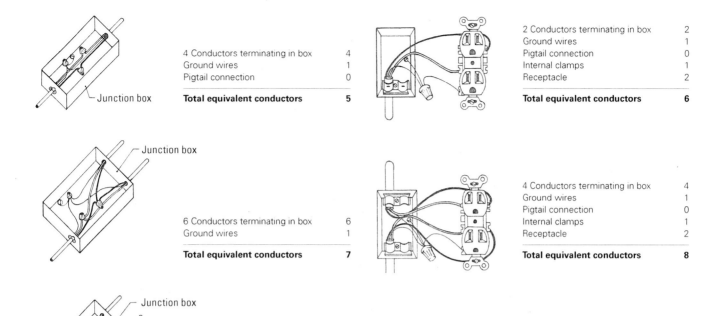

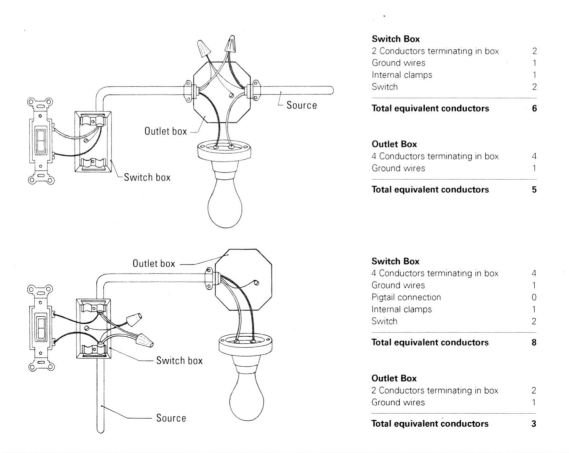

Figure 1-9. Examples of equivalent number of conductors in boxes.

CHAPTER 1. ELECTRICAL DESIGN AND COMPONENT BASICS

EXAMPLE

1-1. Determining junction box size.

Determine the required size of a junction box that has four cables entering it as shown in the third example in the first column of Figure 1-9. Each cable has two AWG (American Wire Gage) 12 conductors and a ground.

SOLUTION

From Figure 1-9, this box has nine equivalent conductors. From Table 1-3, AWG 12 conductors require 2.25 cubic inches of box volume per equivalent conductor. Required box volume is:

$$(9 \text{ equivalent conductors})(2.25 \text{ in}^3 \text{ per equivalent conductor}) = 20.25 \text{ cubic inches}$$

From Table 1-3, a metal box would have to be at least 4- x 4- x 1.5-inches. If the box is plastic, obtain one with a volume larger than 20.25 cubic inches stamped on the inside.

Plastic conduit

Because steel conduit corrodes in **Damp** buildings, the most common conduit for **Damp** buildings is Schedule 40 PVC (polyvinyl chloride) plastic. All fittings and boxes for PVC conduit must be noncorrosive. PVC conduit is available in 10- and 20-foot lengths and in diameters of 1/2 inch to 6 inches.

Use conductors with a type W designation—**THWN**, **RHW**, **THHW** and **XHHW**, *NEC 31010(C)*. Table 1-4 lists the classifications of conductor insulation. Use bare or green-covered copper conductors for the bonding (grounding) wires in PVC conduit. See Table 1-5 for the maximum number of wires per conduit for 1/2-, 3/4-, and 1-inch sizes.

Surface mount conduit on walls and ceilings. Avoid running the conduit inside building walls or above ceilings. If conduit must pass through a wall or ceiling from a warm to a cold area, seal the inside of the conduit with press-in-place putty on the warm side of the wall or ceiling, *NEC 300.7(A)*. This putty is called duct sealer and is available from electrical supply stores. The seal prevents moisture in the warm part of the conduit from migrating into the cold part and condensing.

Table 1-3. Metal box volume and equivalent conductor capacity.
Based on *NEC Table 314.16(A)* Equivalent conductors include devices as well as conductors that are in the box. See text for discussion.
Note: Plastic boxes are stamped with volume and some have maximum equivalent conductors.

Box dimensions (inches)	Box volume* (cubic inches)	Maximum equivalent conductor per box			
		Conductor size (AWG):			
		14	12	10	8
4 x 1.25 (round or octagonal)	12.5	6	5	5	4
4 x 1.50 (round or octagonal)	15.5	7	6	6	5
4 x 2.125 (round or octagonal)	21.5	10	9	8	7
4 x 4 x 1.25	18.0	9	8	7	6
4 x 4 x 1.50	21.0	10	9	8	7
4 x 4 x 2.125	30.3	15	13	12	10
4.6875 x 4.6875 x 1.25	25.5	12	11	10	8
4.6875 x 4.6875 x 1.50	29.5	14	13	11	9
4.6875 x 4.6875 x 2.125	42.0	21	18	16	14
3 x 2 x 1.50 device	7.5	3	3	3	2
3 x 2 x 2 device	10.0	5	4	4	3
3 x 2 x 2.25 device	10.5	5	4	4	3
3 x 2 x 2.50 device	12.5	6	5	5	4
3 x 2 x 2.75 device	14.0	7	6	5	4
3 x 2 x 3.50 device	18.0	9	8	7	6
4 x 2.125 x 1.50 device	10.3	5	4	4	3
4 x 2.125 x 1.875 device	13.0	6	5	5	4
4 x 2.125 x 2.125 device	14.5	7	6	5	4

* Box volumes are **usable volumes** and will not equal calculated volumes based on box dimensions.

Table 1-4. Conductor insulation classifications.

Insulations material	Insulation type (letter)	Description
Rubber	RHW	Moisture-resistant and flame-retardant thermoset
Thermoplastic	THHW	Flame-retardant, moisture-and heat-resistant thermoplastic
	THWN	Flame-retardant, moisture-and heat-resistant, with nylon jacket outer covering
Cross-linked synthetic polymer	XHHW	Flame-retardant and moisture-resistant thermoset
Hard service cord	SEO	Thermoplastic elastomer, oil-resistant (sunlight-resistant and flexible in cold weather)
	SOOW	Oil-, heat-, moisture- and ozone-resistant (extra flexible from 221° F to 58° F
	SO	Thermoset insulated with oil-resistant thermoset cover, no fabric braid
	STO	Thermoplastic or thermoset insulated with oil-resistant thermoplastic cover, no fabric braid

Table 1-5. Maximum number of conductors in Schedule 40 rigid PVC conduit.
Based on *NEC Annex Table C, 10*.

Insulation type	Conductor size (AWG)	Conduit size 1/2"	Conduit size 3/4"	Conduit size 1"
XHHW	14	8	14	24
	12	6	11	18
	10	4	8	13
	8	2	4	7
RHW, RHW-2	14	4	7	11
	12	3	5	9
	10	2	4	7
	8	1	2	4
THWN, THHN	14	11	21	34
	12	8	15	25
	10	5	9	15
	8	3	5	9

a. Coupling   b. Terminal adaptor   c. Female adaptor

Figure 1-10. PVC couplings and threaded adaptors.

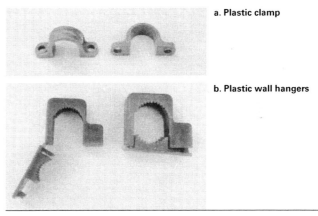

a. Plastic clamp

b. Plastic wall hangers

Figure 1-11. Fasteners for PVC conduit.

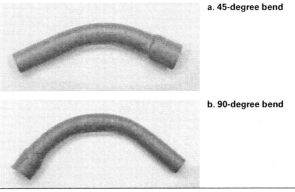

a. 45-degree bend

b. 90-degree bend

Figure 1-12. Premade PVC conduit bends.

Cut PVC conduit with a fine-toothed saw or a special plier-type cutter. The cutter is faster and makes a smoother cut. After cutting, ream saw cuts smooth with a file. Permanent joints are made with PVC couplings and are solvent-welded with a solvent-cement. For nonpermanent joints, use threaded adapters sealed with rubber washers, as depicted in Figure 1-10. The threaded adapters are solvent-welded to the conduit, but can be removed from the boxes.

Support 1/2- to 1-inch diameter PVC conduit at least 3 feet on-center with nonmetal fasteners, *NEC 352.30(B)*. Figure 1-11 shows fasteners for PVC conduit. Where walls are washed frequently, space the conduit, boxes and fittings at least 1/4 inch out from walls, *NEC 300.6(D)*. You can buy PVC conduit elbows and offsets, like those shown in Figure 1-12, or you can heat the conduit and bend it by hand. Heat the conduit with a hot box or a hot air blower—never use an open flame. Maintain the circular cross section of the conduit through the bend. Put no more than the equivalent of four 90-degree bends between pull points, for example, conduit bodies and boxes, *NEC 352.24*.

Allow for thermal expansion in PVC conduit by leaving the bends unrestrained. Figure 1-13 illustrates this concept. If the conduit has few corners or is exposed to a wide temperature range and is 40 feet or more in length, it must be made expandable *NEC 352.44*. Expansion joints like those in Figure 1-13 are made for this purpose. See Table 1-6 for change in length of PVC conduit for various temperature changes.

Use PVC molded junction and outlet boxes as shown in Figure 1-6. Connect conduit to boxes as shown in Figure 1-14, or solvent-weld directly into boxes that are manufactured with socket fittings.

Flexible connections to motors and equipment must be moisture-tight, dustproof, and corrosion-resistant. Use nonmetallic, liquid-tight, flexible conduit, as depicted in Figure 1-15, or liquid-tight, plastic-covered, flexible metal conduit—ordinary flexible metal conduit is not suitable. Flexible connections to portable equipment can be hard service cord such as types **SEO**, **SOOW**, **SO** and **STO** instead of conduit. Dustproof and moisture-tight plugs and caps are required for locations such as feed rooms, grain dryers, and silos.

a. Unrestrained bends

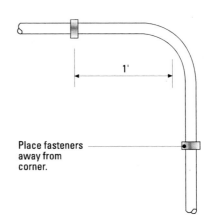

b. Expansion joints

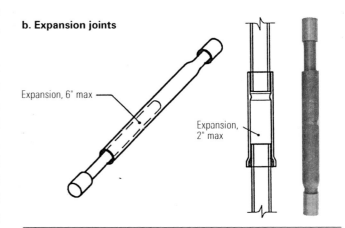

Figure 1-13. Allow for thermal expansion in PVC conduit.
Place fasteners about 1 foot away from bends. If there are only a few bends, install a 6-inch expansion joint every 120 feet or a 2-inch expansion joint every 40 feet.

Table 1-6. Expansion characteristics of PVC rigid nonmetallic conduit.
Coefficient of thermal expansion is 3.38×10^{-5} in/in/°F

Temperature change (F)	Change in length (inches per 100 feet of PVC conduit)
20	0.81
30	1.22
40	1.62
50	2.03
60	2.43
70	2.84
80	3.24
90	3.65
100	4.06
110	4.46
120	4.87
130	5.27
140	5.68
150	6.08
160	6.49

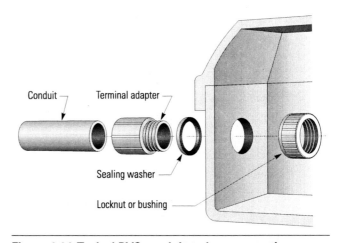

Figure 1-14. Typical PVC conduit-to-box connection.
Place washer over the threads of the terminal adapter. Insert the adapter threads through knock-out and fasten with a standard locknut or threaded bushing.

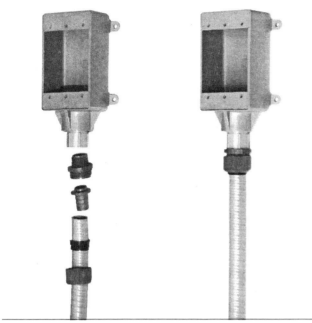

Figure 1-15. Liquid-tight flexible nonmetallic conduit.

Dusty Buildings

Commercial **Dusty** buildings, such as grain elevators, that can have relatively high levels of explosive dust require dust-ignition-proof materials and wiring techniques, *NEC 502.10(A)*. (Hazardous Location, Class II, Division I). These structures are not covered in this handbook.

Farm grain-feed centers are usually not considered to be **Dusty** buildings by local codes, and **Damp** building wiring techniques can be acceptable. However, if high dust levels are anticipated, wiring them as **Dusty** buildings is prudent. Wiring materials and techniques described in *NEC 502.10(A)* would be appropriate.

Planning

Before wiring a building, draw plans locating receptacles, lights, boxes, switches, motors, and all other electrical equipment. Show the route, size, and junctions of all conductors. Show each circuit and the equipment on it. This plan aids in preparing a list of materials required for the electrical system and in locating all the electrical equipment properly. Typical plans are in **Chapter 8, Example Buildings**. Figure 1-16 lists symbols for common electrical equipment.

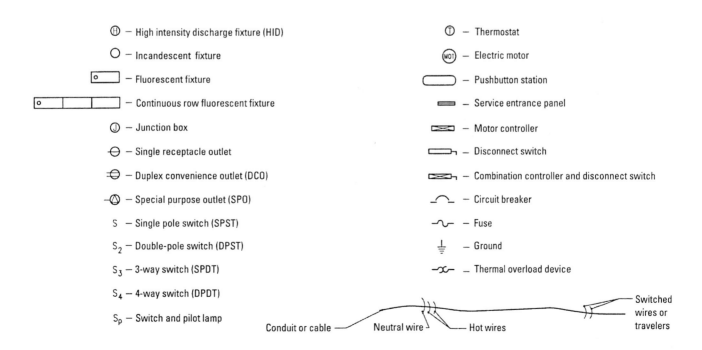

Figure 1-16. Electrical equipment symbols.

CHAPTER 1. ELECTRICAL DESIGN AND COMPONENT BASICS

Lights

The most common types of agricultural lights are
- Incandescent-type.
- Fluorescent (regular and compact).
- High intensity discharge (HID).

Each type has individual properties of light output, maintenance, color, efficiency, and cost that affect selection for a particular task. Consider fluorescents for mounting heights below 10 feet. For 10- to 20-foot mounting heights, consider fluorescents or low-bay HID fixtures. Above mounting heights of 20 feet, use hi-bay HID fixtures.

Table 1-7 lists the five light characteristics that can be used to compare light performance: power use, total light emitted, efficiency, color rendering index (CRI), and lifespan. Power use is the amount of electricity usage of a bulb, which is measured in watts. Total light emitted is the amount of light produced by a bulb. It is measured in lumens. The efficiency of a light bulb is measured in lumens of light produced divided by the watts of electricity used. The CRI measures the effect a light source has on perceived color of an object. A high CRI value means a bulb makes all colors look natural; a low CRI value means a light will cause some colors to wash out or appear to be a completely different hue. The lifespan of a bulb is an estimated number of hours a light bulb will last.

As of this printing, production of some incandescent bulbs is being phased out by government decree. The first to go are the 100W bulbs in October 2012, followed by the 75W in 2013, with the 60W and 40W ending production in 2014. Since these bulbs will no longer be manufactured, the supply of these will be very limited in the future.

At present, some manufacturers are producing halogen incandescent replacement bulbs. These bulbs are slightly more efficient, have a high CRI, and have a longer life than standard incandescents. An example is shown in Figure 1-17.

Incandescent

Consider incandescent-type fixtures when light is needed for short periods and when lights are turned on and off frequently. Figure 1-17 shows a common incandescent bulb. These bulbs reach full output almost immediately. Their initial cost is relatively low, and they operate well in most conditions, including low temperatures.

Bulbs up through 300 watts have standard screw bases that fit ordinary medium-base sockets. Porcelain sockets withstand high heat levels and are recommended for incandescent-type lights over 100 watts.

Incandescent and incandescent replacement bulbs have light efficiencies that are relatively low, so they are the most expensive to operate. In Table 1-7, note that their efficiency increases with wattage. Standard incandescent lights are short-lived—only 750 hours for 100- to 150-watt bulbs. Light output decreases from 80% to 90% of initial value as the bulb approaches its rated life.

A 100-watt incandescent bulb radiates 10% of the input energy as visible light and 72% in the infrared spectrum, accounting for the low efficiency. The high infrared output can also be detrimental in some applications. Incandescent lights have the highest CRI value so should be used in areas where determining natural colors is important. Halogen incandescent replacement bulbs emit less infrared and more UV. The enclosures are coated to reduce the amount of UV released.

Fluorescent

Fluorescent fixtures cost more than incandescent fixtures but produce 3 to 4 times more light per watt, as shown in Table 1-7. Turning these lights on and off frequently or leaving them on for only 2 to 3 hours at a time reduces lamp life. Near the end of its life, a typical fluorescent lamp emits only 60% to 80% of its initial light output.

Because they are temperature sensitive, fluorescent lights are used mainly indoors. Standard indoor lights perform acceptably down to 50 °F; many manufacturers now offer lamps that start and perform well at 0°F. Special ballasts for use down to -20 °F are available. High relative humidity

a. Common   b. Halogen

Figure 1-17. Incandescent bulbs.

Table 1-7. Characteristics of lights.
HPS lamps operate much longer than shown but with greatly reduced output. Total amount of light emitted is equal to the average lumens. Efficiency (lumens per watt) is based on converting electricity into light.

Lamp (watts)	Rated size (watt)	Power use (watt)*	Total light emitted (lumens)	Efficiency (lumens/watt)*	CRI	Lifespan (hours)
Incandescent— Standard (halogen replacement)						
	40	40 (28)	410 (340)	10 (12)	100	1,500 (2000)
	60	60 (43)	780 (750)	13 (16)	100	1,000 (2000)
	100	100 (70)	1,580 (1300)	16 (19)	100	750 (2000)
	150	150 (100)	2,500 (1800)	17 (18)	100	750 (2000)
	300	300 (NA)	5,860	20	100	750
Fluorescent — (cool white or warm white)						
Compact						
Spiral, T3	26	26	1,700	65	82	8,000
Spiral, T4	42	42	2,650	63	82	10,000
Biax, T4	20	20	1,200	60	82	12,000
	28	28	1,750	63	82	12,000
Tube (T8)						
24-inch	17	19	1,325	70	75	20,000
48-inch	32	35	2,800	80	82	20,000
	44	50	4,000	80	86	18,000
60-inch	55	61	5,050	83	86	18,000
96-inch	51	56	4,000	71	75	7,500
	86	95	8,200	86	86	24,000
High intensity discharge (HID)						
Metal halide (MH)						
17	50	60	3,450	58	70	15,000
	70	80	5,200	65	75	15,000
	100	115	8,500	74	75	15,000
	150	175	12,900	74	75	15,000
28	175	210	15,000	72	65	7,500
	250	290	15,000	52	65	10,000
	400	460	40,000	87	65	20,000
High-pressure sodium (HPS)						
17	50	70	4,000	57	22	24,000
	70	90	6,300	70	22	24,000
	70	80	3,800	48	65	10,000
	100	125	9,500	76	22	24,000
	150	180	15,800	88	22	24,000
	150	170	10,500	62	65	15,000
	200	240	22,000	92	22	24,000
	250	295	29,000	98	22	24,000
	310	365	37,000	101	22	24,000
	400	470	50,000	106	22	24,000
18	250	300	22,500	75	85	15,000

*Including the ballast for fluorescent and HID fixtures.

alters the electrostatic charge on the outside of fluorescent tubes, making them difficult to start. Problems develop above 65% relative humidity and become severe at 100%. Some fluorescent bulbs are available with a silicone coating to ensure starting at any relative humidity. Figure 1-18 displays a common fluorescent tube lamp. There are four standard colors of fluorescent tube lamps:

- Standard cool white.
- Deluxe cool white.
- Standard warm white.
- Deluxe warm white.

Standard whites give the highest light output, but they are not desirable for color-matching tasks. Deluxe whites are about 25% less efficient than the standard. Fluorescents have high CRI characteristics; however, some of them may produce light more in the blue light spectrum, which enhances the colors in the green, blue, violet, and magenta range. The deluxe cool white produces light that is most like daylight. The deluxe warm white produces light that is closest in color to the incandescent. Fluorescents produce little infrared light.

Figure 1-19 shows the compact fluorescents(CFL) that are available to screw into incandescent light fixtures. They are available in several wattages. They are energy efficient, saving up to 70% versus incandescent. CFL's are more expensive than incandescent bulbs to purchase, but can last up to 16 times longer because of their lower power usage. CFL's may not hold up to power surges well, so use them in areas where power surges are not a concern. Today's CFL's have very high CRI values so they are closer in performance to incandescent. A drawback of CFL's is the delay in reaching rated light output.

T-8 and T-5 lamps are generally superior to the T-12 technology based upon color rendering and efficiency. A T-8 system with magnetic ballasts will provide 8 to 9% more light and use 4 to 10% less power than a system using T-12 lamps with equal color rendering. Manufacturing of T-12 lamps is being discontinued.

High intensity discharge (HID)

HID lamps include mercury, metal halide and high-pressure sodium; Figure 1-20 shows an example of an HID fixture. They tend to have long lives and to be very energy efficient. They operate well in cold temperatures. Light output is colored; for example mercury is greenish-blue, and sodium is golden yellow. Metal halide lamps are now more popular for many tasks because of their good color

Figure 1-18. Fluorescent tube.

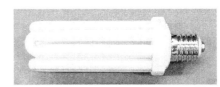

a. Biax

b. Energy smart   c. Spiral

Figure 1-19. Compact fluorescent bulbs.

Figure 1-20. High intensity discharge fixture.

rendering. Many high-pressure sodium lamps have low CRI values, but there are some lamps available that have higher CRI values and are not yellow in output.

HID lamps require 5 to 15 minutes to start and are not usable where lamps are turned on and off frequently. HID lamps are best where mounted at least 12 feet high and where lamps are on for at least 3 hours. Common uses include feedlots, outdoor security lighting, and night lights in dairy freestall barns. Metal halide lamps are often used in dairy parlors. Metal halide lamps have very good CRI values. Mercury vapor lamps are not a good option because they dim over time instead of burning out. Some HID lamps, especially low-pressure sodium, are unsuitable for use where people are working because of their poor color rendering.

Outlets

There are two types of receptacle outlets:
- Duplex convenience outlets (DCO).
- Special purpose outlets (SPO).

Install DCOs where portable equipment (heat lamps, power tools, etc.) is used often. Locate additional DCOs along alleys, near doors, etc., for convenient short-term use of portable equipment. Avoid long-term use of extension cords. Most DCOs in livestock buildings now require ground-fault-circuit interrupter (GFCI) protection, *NEC 547.5(G)*.

SPOs are for motors and equipment that are on individual branch circuits. An SPO can have a receptacle for plug-in equipment, or the equipment can be hard-wired (no plug and receptacle) into the outlet box.

Install SPOs next to the fixed equipment that they serve. Fixed equipment stays in one place for long periods of time (for example, feed handling equipment, ventilating fans, heat lamps, tank heaters, and gutter cleaners).

Electrical Layout Guidelines

This section outlines the guidelines for locating lights, duplex convenience, and special purpose outlets. These guidelines are based on many years of experience by designers and engineers, and are based on providing maximum convenience and performance.

Lights

Use enough lighting for efficient inspection and work. Provide at least two lighting circuits in each building. Some areas, such as animal pens, may require two light-intensity levels—consider two rows of lights on separate switches.

Install enough switches so at least one light in each room is controlled from main access doors. Consider a separate switch for each row of lights and for lights over feeding stations for all-night feeding. Equip outdoor and haymow light switches with pilot lights to indicate when lights are on.

No light reflectors are necessary when ceiling and wall surfaces are white. Provide reflectors if surfaces are dark.

Table 1-8 shows illumination levels and light energy requirements for agricultural-type buildings using fluorescent or incandescent-type sources. Table 1-9 shows illumination levels and light energy requirements for HID sources. Table 1-10 shows typical illumination levels that can be used to gain better perspective on lighting levels in a room. See **Chapter 8, Example Buildings**, for more information on using Tables 1-8 and 1-9.

Consider ease of bulb replacement and problems with bugs, birds, and mechanical damage when locating lights. Also, consider shadows, glare, and unwanted light. Table 1-11 shows general guidelines for locating lights in agricultural-type buildings. Figure 1-21 shows fluorescent and CFL lights being used in a farrowing room.

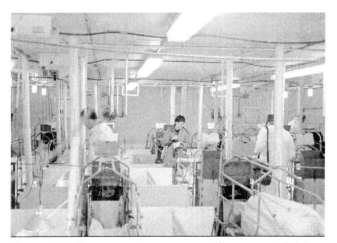

Figure 1-21. Fluorescent and compact fluorescent lights used in a farrowing room.

Duplex convenience and special purpose outlets

Locate DCOs at least 4 feet high—higher if they may be damaged by animals. Table 1-12 shows general guidelines for locating duplex convenience outlets in agricultural-type buildings.

Table 1-8. Interior illumination levels and light energy requirements for agricultural-type buildings.

Space lights uniformly throughout the work area. Some areas may need additional task lighting, for example, the office. These table values assume frequent cleaning of bulbs (once a month in animal buildings) and replacing bulbs at rated life. With infrequent cleaning and delayed replacement, increase table values 40%. Dairy, poultry, and general illumination level recommendations are from ASABE, which are comparable to other organizations' recommendations.

This table assumes an 8-foot high ceiling, and 70% ceiling and 50% wall reflectance. For lights located higher in rooms with low reflection surfaces, consider dome reflectors and higher wattage incandescent bulbs. Incandescent: maintenance factor (MF) = 0.75 and coefficient of utilization (CU) = 0.69. Fluorescent: MF = 0.70 and CU = 0.66.

| | Level of illumination | | Standard cool white fluorescent | Standard incandescent* | | |
| | | | 40 W | 100 W | 150 W | 300 W |
Task	(lux)	(foot-candles)	(watts per square foot)			
Dairy						
Cow housing	75	7	0.19	0.80	0.70	0.64
Calf housing	110	10	0.28	1.14	1.00	0.91
General milking operation	215	20	0.55	2.29	2.00	1.82
Equipment washing area	1,075	100	2.75	11.4	10.0	9.10
Milk handling	215	20	0.55	2.29	2.00	1.82
Parlor, pit and near udder	540	50	1.38	5.72	5.00	4.55
Loading platform	215	20	0.55	2.29	2.00	1.82
Poultry						
Brooding and laying hens	215	20	0.55	2.29	2.00	1.82
Egg handling	540	50	1.38	5.72	5.00	4.55
Egg processing	750	70	1.93	8.01	7.00	6.37
Swine						
Farrowing	110	10	0.28	1.14	1.00	0.91
Nursery	55	5	0.14	0.57	0.50	0.46
Growing/finishing	55	5	0.14	0.57	0.50	0.46
Breeding/gestation	160	15	0.41	1.72	1.50	1.37
Animal inspection/handling	215	20	0.55	2.29	2.00	1.82
Sheep						
Lambing	160	15	0.41	1.72	1.50	1.37
Growing/finishing	110	10	0.28	1.14	1.00	0.91
Animal inspection/handling	215	20	0.55	2.29	2.00	1.82
Beef						
Housing	55	5	0.14	0.57	0.50	0.46
Along feedbunk	110	10	0.28	1.14	1.00	0.91
Animal inspection/handling	215	20	0.55	2.29	2.00	1.82
Equine						
Alley	430	40	1.10	4.58	4.00	3.64
Arena	430	40	1.10	4.58	4.00	3.64
AI lab	750	70	1.93	8.01	7.00	6.37
Breeding	215	20	0.55	2.29	2.00	1.82
Box stalls	110	10	0.28	1.14	1.00	0.91
Tack room	430	40	1.10	4.58	4.00	3.64
Vet area	540	50	1.38	5.72	5.00	4.55
Wash rack	430	40	1.10	4.58	4.00	3.64
General						
Feed storage/processing	215	20	0.55	2.29	2.00	1.82
Office, general	540	50	1.38	5.72	5.00	4.55
Office, tasks	1,075	100	2.75	11.4	10.0	9.10

*Use of the halogen incandescent replacement will not appreciably change the watts per square foot calculation.

Table 1-8. Continued.

Task	Level of illumination		Standard cool white fluorescent	Standard incandescent*		
			40 W	100 W	150 W	300 W
	(lux)	(foot-candles)	(watts per square foot)			
General (continued)						
Haymow and silo	30	3	0.08	0.34	0.30	0.27
Machine storage	55	5	0.14	0.57	0.50	0.46
Machine repair area	320	30	0.83	3.43	3.00	2.73
Shop bench area	1,075	100	2.75	11.4	10.0	9.10
Rough bench work	540	50	1.38	5.72	5.00	4.55
Restrooms	320	30	0.83	3.43	3.00	2.73
Storage, active	110	10	0.28	1.14	1.00	0.91
Storage, general	55	5	0.14	0.57	0.50	0.46

*Use of the halogen incandescent replacement will not appreciably change the watts per square foot calculation.

Table 1-9. Interior light energy of high-intensity discharge lights for agricultural-type buildings.

Level of illumination		Metal Halide	High-pressure sodium		
			50 W	70 W	100 W
(lux)	(foot-candles)	70 W	(watts per square foot)*		
30	3	0.07	0.06	0.09	0.08
55	5	0.12	0.10	0.15	0.12
75	7	0.18	0.14	0.21	0.18
110	10	0.25	0.20	0.31	0.25
215	20	0.51	0.40	0.61	0.05
320	30	0.77	0.60	0.92	0.75
540	50	1.28	1.00	1.53	1.25
1,075	100	2.57	2.00	3.07	2.5

*Based on lights that are 20 feet above a surface, and a 20- x 20-foot illuminated area, Light loss factor (LLF) = 0.75 and Lamp lumen depreciation factor (LLD) = 0.75.

Table 1-10. Typical illumination levels.
Table is intended to be used to help provide a baseline perspective when estimating a room's illumination level.

Comparison item	Level of illumination	
	(lux)	(foot-candles)
Moonlight	0.1	0.01
Dimly lit restaurant	10	1.0
Bright room	1,075	100
Bright sunlight	107,500	10,000

Table 1-11. Guidelines for locating lights in agricultural-type buildings.

Building type	Guideline
Beef housing	Provide one row of lights over feedbunks. Install additional lights in pens or handling areas, especially for veterinary care. Place lights to aid chores, brighten dark corners and eliminate long shadows.
Dairy, Stanchion barn*	Install lights about 12 feet on-center along centerline of feed alleys. Face out arrangements: place incandescent lights along litter alley centerline; one light directly behind every other stall divider, or install a continuous row of two lamp fluorescent lights along alley centerline for even lighting of cow udders. Face in arrangements: place light fixtures about 1 foot to rear of gutter and directly behind every other stall divider.

*Contact your dairy inspector concerning regulations.

Continued

Table 1-11. Continued.

Building type	Guideline
Dairy, Freestall barn*	Place a row of lights over center of each work/litter alley and over feedbunks. Space at about 2 times light height. Add lights to aid specific tasks or brighten dark corners and avoid long dark shadows.
Dairy, Pens and stalls*	Provide one light fixture per bull, maternity or calf pen. Locate switch outside of pen. Provide one row of lights over each row of calf stalls.
Dairy, Milking parlor*	Provide one row of lights along each side of the operator's pit. Locate lights so operator does not work in any shadows. Install one light at each animal entrance.
Dairy, Milk room*	Provide one fluorescent light over the outer edge of each wash vat and an additional fluorescent light for each 100 square feet of floor area. Use fixtures with two 4-foot long, T8 32-watt cool white fluorescent tubes. Locate fixtures away from bulk tank openings to keep broken glass out of the tank.
Farm shop	Two 150-watt halogen fixtures about 50 inches above a bench and 4 feet apart provide good lighting. For continuous lighting, use fluorescents above benches. Position lamps above the front third of the bench. Provide lights over each concentrated work area, such as welding.
Poultry Housing, Layers	In general, space lights 8 to 12 feet on-center along building width and 10 to 16 feet on-center along the building length, with end fixtures 4 to 5 feet from walls. Refer to bulletins from your local agricultural college or power supplier for more specific information for your region and operation type.
Sheep Housing	Locate lights over pens and feeders to enhance inspection. Provide one light 12 to 15 feet on-center down alley centerline along lambing or feeding pens.
Swine Housing	Provide one row of lights over each row of pens or stalls and one row along feed alley center. Install at least one light over every other pen partition.

*Contact your dairy inspector concerning regulations.

Table 1-12. Guidelines for locating duplex convenience outlets (DCO) and special purpose outlets (SPO) in agricultural-type buildings.

Building type	Guideline
Beef	Provide one DCO wherever portable equipment is likely to be used. Install one near each major entrance to the building.
Dairy, Stanchion barn*	Provide DCOs 20 feet on-center along litter alley. They can be on outside walls where cows face in. Install extra ones in feed alley for supplemental lighting.
Dairy, Freestall barn*	Provide DCOs wherever portable equipment (tools, circulation fans, spot heaters, etc.) may be used. Also provide one on inside wall near each major entrance. Provide SPOs for 3-phase circulation fans.
Dairy, Pens and stalls*	Provide one DCO for each pen for equipment such as heat lamps or groomers. In stall areas, install at least one on each wall and no more than 20 feet apart.
Dairy, Milking parlor*	Provide one DCO at each end of operator's work area. Add SPOs for heating equipment.
Dairy, Milk room*	Provide one DCO for each work station. DCOs for radio, clock and desk lamp may be desirable for recordkeeping room. A receptacle SPO on a manual switch may be required for electric pump on milk truck; check local regulations for location, type and rating. Place outlets about 4 feet high to avoid splashed water.
Farm shops	Provide one DCO every 4 feet along work bench, preferably under the front edge to keep cords off the bench. Provide one DCO 10 feet on-center and 4 feet off the floor along other shop walls. Locate one SPO at each permanent location of a motor driven device.
Machine storage buildings	Provide one DCO per 40 feet of wall perimeter.
Poultry	Provide DCOs for tools, water warming, special lighting, etc. Install one per 400 ft2 of floor area with at least one per pen.
Sheep	Provide one DCO per two stalls or pens. Provide one DCO 15 to 20 feet on-center along alleys. Provide extras near water troughs (for water warming), near shearing stations and at each major building entrance.
Swine	Provide at least one DCO per two farrowing stalls or two pens. Locate over heat lamp locations. Provide one DCO 15 to 20 feet on-center along alleys.

*Contact your dairy inspector concerning regulations.

CHAPTER 2

Branch Circuits

BRANCH CIRCUITS carry current from the short-circuit protection device (fuse or circuit breaker) to the loads (lights, outlets, motors). They are called 120- or 240-volt circuits. Actual voltages generally vary from 108 to 125 volts for nominal 120 volts and from 220 to 250 volts for nominal 240 volts.

These circuits have wires for conducting current and for grounding. There are two types of grounding:
- **System grounding** is the grounding of one of the current-carrying conductors at the service entrance panel (SEP). The grounded conductors are called **circuit neutrals**. These conductors normally carry current when 120-volt loads are operating.
- **Equipment grounding** is the grounding or bonding of noncurrent-carrying equipment such as motor frames back to the service entrance panel grounding bar. The equipment grounding wire serves only to ground the metal frames of electrical equipment. The grounding conductor does not carry current during normal operation, but will carry current in case of damage to or defect in electrical equipment or wiring. The conductors for grounding equipment are called **equipment grounds**. Bonding is used to describe the functions of the equipment grounding conductors. This is being done to better differentiate the differences in connecting to earth (grounded) and connecting (bonding) equipment that is likely to become energized by a ground fault.

An equipment bonding conductor must make a **continuous** connection from the grounding bar in the service entrance panel (SEP) to all receptacles, metal fixtures, and motor frames. Run the equipment grounding (bonding) conductor into all boxes—if the boxes are metal, they must also be grounded. Use only grounding-type receptacles and switches, and connect the equipment grounding conductor to the green grounding screw. Connect the equipment grounding conductor to the frame of metal fixtures or motors.

120-volt Circuits

Circuits at 120 volts have one hot wire, one circuit neutral, and one equipment grounding conductor. The hot wire is usually black or red; the circuit neutral is white, and the equipment grounding conductor is bare or green.

Do **NOT** connect the equipment grounding conductor to the circuit neutral, *NEC 250.142(B)*. Even though these two conductors may be connected to the same busbar in the service entrance panel, a connection in the branch circuit can lead to a dangerous shock hazard and to current in the equipment grounding conductor. For more information, see **Chapter 6, Stray Voltage**.

The circuit neutral and equipment grounding conductor must run without interruption to all 120-volt equipment:
- Do not protect circuit neutrals or equipment grounding conductors with fuses or circuit breakers.
- Do not install switches on circuit neutrals or equipment grounding conductors—switches can be installed on hot wires only, as illustrated in Figure 2-1.

240-volt Circuits

Circuits at 240 volts have two hot wires and an equipment grounding conductor. Circuit neutrals are not required for most 240-volt equipment. If an appliance has both 120- and 240-volt circuits (for example, ranges, clothes dryers), both the circuit neutral and equipment grounding conductor are required.

Three wires run from the power supplier transformer to the building. One of the wires is grounded (connected to

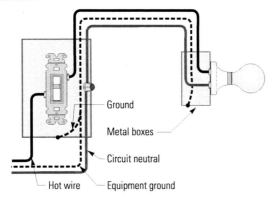

Figure 2-1. Light with one switch.
In 120-volt circuits, white wires may not be hot wires except on switch loops. Note that the switch is installed in the hot wire—not the circuit neutral. Connect the equipment ground to all metal boxes and equipment. In these circuits, the equipment ground conductor is connected to the green grounding screw on the switches. Light fixtures of the type shown are not connected to the equipment ground conductor.

earth) at the transformer and at the service entrance of the building. It is called the neutral (wire N of Figure 2-2). The voltage between conductor **A** and **B** of Figure 2-2 is 240 volts, while the voltage between **A** (or **B**) and N is 120 volts.

Circuit Types

Two types of branch circuits are discussed here:

- **General purpose branch circuits** are for loads such as lights and DCOs. DCOs can be used for portable appliances under 1,500 watts and portable power tools. DCOs are only for small or periodic loads: it is assumed that all the DCOs on one circuit will not have large loads at the same time. For agricultural-type buildings, a general-purpose circuit should have a 120-volt, 20-amp circuit with AWG 12-CU (copper) conductor. Use 20-amp rated materials where rugged service is required. It is recommended that no more than ten DCOs or light fixtures be on one general-purpose circuit. Figure 2-3 illustrates this principle.
- **Individual branch circuits** are for known, specific loads such as stationary motors and appliances and SPOs. An individual circuit may have only one load such as a fan motor, or it may have more than one load such as a number of heat lamps. Individual circuits usually require higher capacity circuit breakers and larger wires than general purpose circuits. Size the individual circuit conductors and

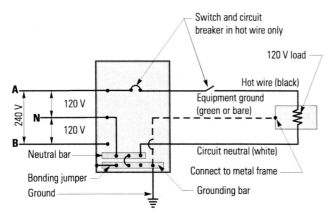
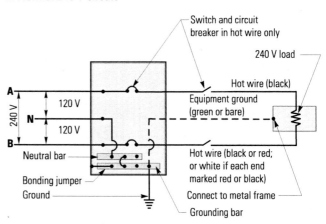

Figure 2-2. Circuit wiring schematics.
With a 240-volt power supply, 120 volts is also available.

20   CHAPTER 2. BRANCH CIRCUITS

circuit breakers for the specific load or loads. These circuits are often on 240-volt service.

Individual circuits and SPOs are recommended for larger motors and heaters. It is recommended that stationary equipment be hard-wired into the SPO. Receptacles and plugs can be used for portable equipment, but use the proper configuration for the voltage and amperage that will be encountered. Figure 2-4 shows typical receptacles and plugs for agricultural-type buildings.

Wire only one or at most two fans per circuit. If possible, install at least two fan circuits in each room of environmentally-controlled animal buildings, so if one circuit fails, a fan on the other circuit can ventilate the room. Use supplementary fusing for each fan so individual fan failure will not trip the main circuit.

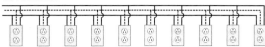

10 duplex convenience outlets (DCO), maximum per 15 A circuit

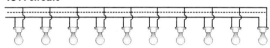

10 - 150 W light fixtures, maximum per 15 A circuit

Figure 2-3. Maximum number of lights or duplex convenience outlets per general-purpose circuit.

For 15-amp, 120-volt circuits. Use 20-amp rated materials where rugged service is required. Average amperage load per duplex convenience outlet (DCO) or light fixture is 1.5 amps (180 watts on a 120-volt circuit). DCOs and lights can be on the same circuit, but turning on a load on a DCO can cause the lights to flicker.

a. Configuration for nonlocking receptacles and plugs

Nominal Voltage	Maximum Amperage							
	15 A		20 A		30 A		50 A	
	Receptacle	Plug	Receptacle	Plug	Receptacle	Plug	Receptacle	Plug
120 V	5-15R	5-15P	5-20R	5-20P	5-30R	5-30P	5-50R	5-50P
240 V	6-15R	6-15P	6-20R	6-20P	6-30R	6-30P	6-50R	6-50P

Note: During receptacle installation, locate the grounding slot for nonlocking plugs at the top as shown. This installation will prevent a slightly dislodged plug from exposing part of the neutral and hot blades. If a metal object should fall and come into simultaneous contact with a plug's partially exposed neutral and hot blades, a short could be created resulting in sparks and possibly a fire.

b. Configuration for locking receptacles and plugs

Nominal Voltage	Maximum Amperage					
	15 A		20 A		30 A	
	Receptacle	Plug	Receptacle	Plug	Receptacle	Plug
120 V	L5-15R	L5-15P	L5-20R	L5-20P	L5-30R	L5-30P
240 V	L6-15R	L6-15P	L6-20R	L6-20P	L6-30R	L6-30P

Figure 2-4. Typical receptacles and plugs for agricultural-type buildings.
National Electrical Manufacturers Association (NEMA) Standard Configuration WD1. Only for 120-volt and 240-volt, two-pole, two-wire circuits with ground.

Motor Circuits

Use totally enclosed motors in **Damp** buildings. Use totally enclosed, air-over cooled motors for ventilating fans. If practical, connect motors to 240-volt service to reduce circuit current and the possibility of stray voltage problems. For several 5-hp or larger motors, consider three-phase service if available.

In this handbook, motor circuits are discussed only for single-phase motors on 120-volt or 240-volt service. The following section discusses only the basics required for motor circuits—there are numerous ways of meeting these criteria, depending on the application. Refer to the current edition of the *NEC* for more details.

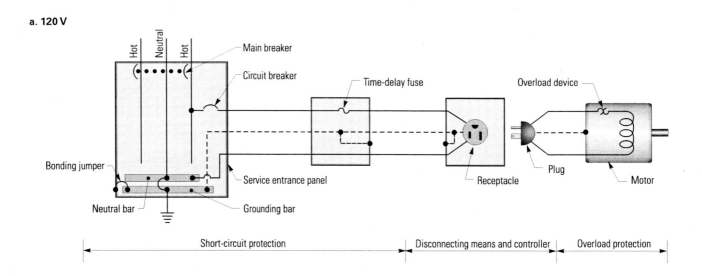

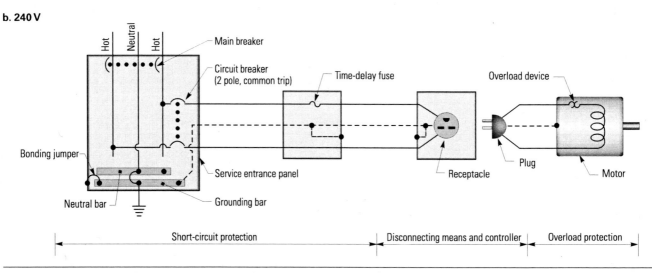

Figure 2-5. A circuit for portable motors one-third horsepower or under.
Receptacle and plug act as the disconnecting means and controller. A receptacle and plug can be used as the controller only for motors one-third horsepower or under. Note that these small motors will generally require a time delay fuse for short-circuit protection because circuit breakers that are small enough to work properly are not easily available.

Every circuit that supplies power to a motor must have the following:
- **Branch circuit short-circuit protection** to protect the circuit conductors against short-circuits and ground-fault currents.
- A **disconnecting means** to completely disconnect the motor from the power supply.
- A **controller** to start and stop the motor. It must be capable of interrupting the stalled motor current.
- An **overload protection device** to interrupt the motor circuit if the motor should fail to start or if overloaded.

See Figures 2-5 to 2-10 and Table 2-1 for examples of common agricultural-type building motor circuits and devices.

a. 120 V

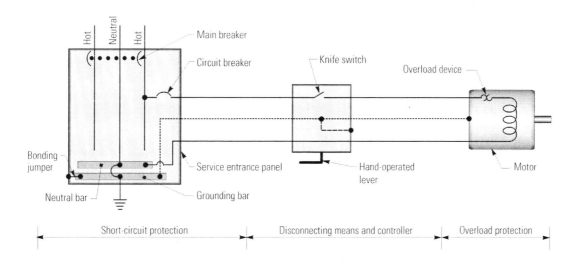

b. 240 V

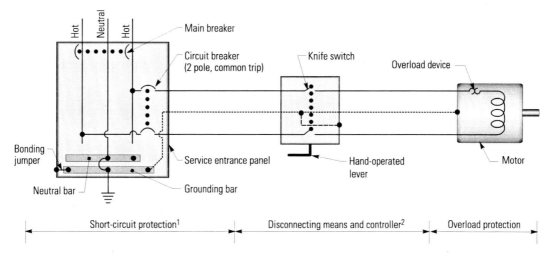

[1] If small enough circuit breakers are not available, a time delay fuse or a supplementary breaker may be required at the disconnecting means for short-circuit protection.

[2] Use a switch rated in horsepower for motors over 2 hp.

Figure 2-6. Motor circuit.
A knife swith for both disconnecting means and controller. The switch must be capable of being controlled by applying the hand to a lever.

a. 120 V

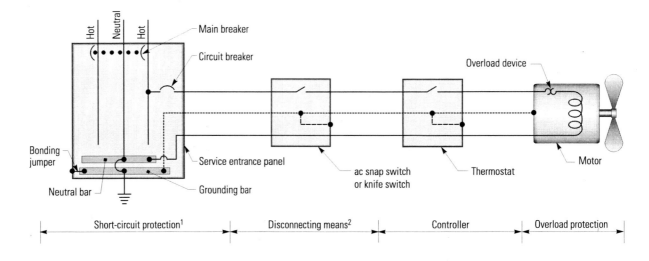

b. 240 V

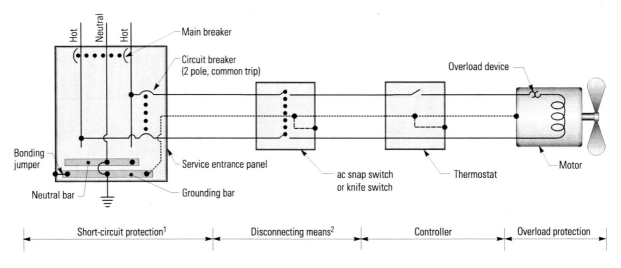

[1] If small enough circuit breakers are not available, a time delay fuse or a supplementary breaker may be required at the disconnecting means for short-circuit protection.

[2] Use a switch rated in horsepower for motors over 2 hp.

Figure 2-7. Ventilating fan motor circuit.
AC snap switch or knife switch is the disconnecting means, and a thermostat is the controller.

a. 120 V

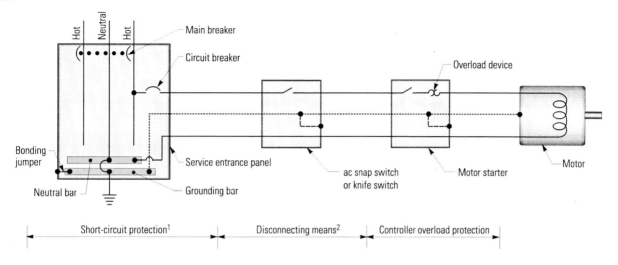

b. 240 V

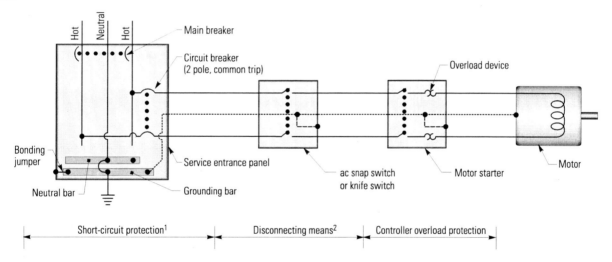

[1] If small enough circuit breakers are not available, a time delay fuse or a supplementary breaker may be required at the disconnecting means for short-circuit protection.

[2] Use a switch rated in horsepower for motors over 2 hp.

Figure 2-8. Motor circuit.
AC snap switch or knife switch is the disconnecting means, and a motor starter is the controller and overload protection.

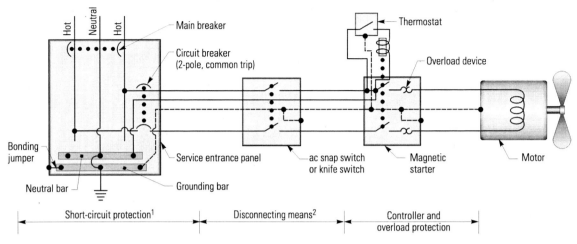

[1] If small enough circuit breakers are not available, a time delay fuse or a supplementary breaker may be required at the disconnecting means for short-circuit protection.

[2] Use a switch rated in horsepower for motors over 2 hp.

Figure 2-9. Ventilating fan motor circuit.
A magnetic motor starter controlled by a thermostat is the motor controller. Use this circuit where the motor current rating is greater than the thermostat current rating. Most thermostats have a current rating of 10 to 20 amps so any 240-volt motor over one horsepower may exceed the thermostat rating.

Table 2-1. Typical equipment for motor circuit devices.
This table lists only common devices used for the function indicated. See the text for determining the proper size or rating required to match the device to the motor.

Function	Common devices
Short-circuit protection Protects circuit against short-circuits and ground-faults.	Inverse time circuit breaker in the SEP
	Time delay fuse with the disconnect switch to protect the motor in addition to a circuit breaker in the service entrance panel to protect the conductors
Disconnecting means Must disconnect both the controller and the motor. Locate within sight and within 50 feet of the motor.	Receptacle and plug (portable motors only)
	AC snap switch rated in amperage (only for motors 2 hp or under and 240 volts or under)
	Knife switch rated in amperage (only for motors 2 hp or under and 240 volts or under)
	Motor circuit switch rated in horsepower
Controller Locate the disconnecting means within sight and within 50 feet of the controller. Cannot be the same device as the disconnect (except under certain conditions as described in text).	Receptacle and plug (only for motors 1/3 hp or under).
	AC snap switch (only for motors 2 hp or under and 240 volts or under)
	Knife switch (only for motors 2 hp or under and 240 volts or under)
	Manual motor starter
	Magnetic motor starter
	Thermostat
	Variable speed controller
	Timer
Overload protection Protects motor against overloads.	Built into motor
	Time delay fuse with the disconnect switch (cannot also serve as the short-circuit device except as described in text)
	Overload device with manual or magnetic motor starter

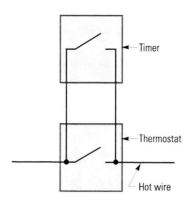

Figure 2-10. Timer wired in parallel with thermostat.
The timer ensures that the fan will run at least a few minutes out of a period of time, typically every 10 minutes. The thermostat makes the fan run continuously at warm temperatures. Use on either 120-volt or 240-volt circuits.

Short-Circuit Protection

The motor branch circuit must have a fuse or circuit breaker to protect wiring against short-circuits and ground-faults. A motor uses more current when starting than when running, so the fuse or circuit breaker protecting the motor branch circuit must have an ampere rating somewhat higher than the motor full-load current to permit the motor to start and accelerate its load, *NEC 430.52*. Circuit breakers are tested to withstand six times their rating for up to 10 seconds to permit motor starting.

For single-phase and polyphase motors, use inverse time breakers—not instantaneous trip breakers. Table 2-2 contains specific information about short-circuit and ground-fault protection. Common circuit breakers are inverse-time breakers. For fuses, use only the time-delay type as shown in Table 2-2. Use a circuit breaker or fuse with an amperage rating large enough to carry the starting current of the motor. There are maximum limits to the size of breaker or fuse that can be used. The maximum size for a time-delay type fuse is 1.75 times the full-load motor current. The maximum size for an inverse time breaker is 2.50 times the full-load motor current. Use the smallest rating that permits the motor to start and operate properly. This may be a trial and error process; it is suggested that you start with a circuit breaker or fuse rated at about 25% over the motor full-load current and if that does not work, try the next larger size. This procedure may result in a circuit breaker or fuse with a rating higher than the conductor ampacity. This is allowed under these circumstances because the conductors are protected from overload by the motor overload device, *NEC 430.52*.

For sizing short-circuit protection, use Table 2-3 to get full-load currents even if the motor nameplate lists current draw. Table 2-3 is only for single-phase ac motors that are not low speed, high torque, multi-speed, or hermetically sealed.

For smaller motors, the maximum limitation on motor circuit breakers may require a breaker smaller than what is readily available. In this case, a time-delay fuse can be used at the motor disconnect in addition to the circuit breaker at the service entrance panel (SEP). Figure 2-11 illustrates a typical method for short-circuit protection for smaller motors.

Table 2-2. Maximum rating of motor short-circuit and ground-fault protection device.
Based on *NEC Table 430.52*.

Motor type	Percentage of full load current		
	Nontime delay fuse [a]	Dual element (Time-Delay) fuse [a]	Inverse time breaker [b]
Single-phase motors	300	175	250
AC polyphase motors other than wound-rotor			
Squirrel cage—other than Design B energy efficient	300	175	250
Design B energy efficient	300	175	250
Synchronous [c]	300	175	250
Wound rotor	150	150	150
Direct current (constant voltage)	150	150	150

[a] The values in nontime delay fuse column apply to time-delay Class CC fuses.
[b] The values given in the last column also cover the ratings of nonadjustable inverse time types of circuit breakers that may be modified.
[c] Synchronous motors of the low-torque, low-speed type (usually 450 rpm or lower), such as those used to drive reciprocating compressors, pumps, and so forth, that start unloaded, do not require a fuse rating or circuit-breaker setting in excess of 200% of full-load current.

Table 2-3. Full-load currents for single-phase AC motors (amps).
These values are for motors running at usual speeds with normal torque characteristics. Do NOT use this table for low speed, multi-speed, high torque or hermetically sealed motors. Based on NEC Table 430.248.

Motor (hp)	Full load current (amps)	
	120 V	240 V
1/6	4.4	2.2
1/4	5.8	2.9
1/3	7.2	3.6
1/2	9.8	4.9
3/4	13.8	6.9
1	16.0	8.0
1-1/2	20.0	10.0
2	24.0	12.0
3	34.0	17.0
5	56	28.0
7-1/2	80	40.0
10	100	50.0

Generally only one motor is installed per branch circuit. More than one motor may be installed per branch circuit under the following conditions:

- Each motor has its own short-circuit protection device in addition to the circuit breaker at the SEP. These individual short-circuit protection devices are typically time-delay fuses at each motor's disconnect.
- Two or more motors (each not over 1 hp and each having a full-load current not over 6 amps) may be placed on one circuit, if the circuit is not protected at over 20 amps for 120 volts, or 15 amps for 240-volt service, *NEC 430.53*.
- Each motor must have overload protection.

Disconnecting Means

Each motor must have an individual disconnecting means, except when a number of motors drive several parts of one machine or system, *NEC 430.112*. The disconnecting means must be capable of disconnecting the motor **and** controller from all ungrounded wires (that is, all hot wires) to protect anyone working on the controller or motor, *NEC 430.101*.

The disconnecting means must clearly indicate whether it is on or off, *NEC 430.104*. Locate the disconnecting means within sight and within 50 feet of the controller and the motor, *NEC 430.102*. No individual poles shall be able to operate independently, *NEC 430.103*.

A motor circuit switch can be the disconnecting means for any motor circuit, *NEC 430.109*. A motor circuit switch is any switch that is rated in horsepower and is sized as follows:

- Switch horsepower rating greater than or equal to motor horsepower rating.
- Switch must also have a current rating greater than or equal to 1.15 times the motor full-load current rating, *NEC 430.110*. Use motor full-load current ratings from Table 2-3.

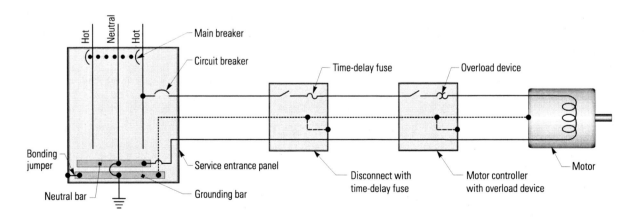

Figure 2-11. Typical short-circuit protection for smaller motors.
For smaller motors, a time-delay fuse at the disconnect may be required in addition to the circuit breaker in the SEP, because such small inverse time breakers are not available.

For motors that are **not** permanently fixed into place, a receptacle and plug can serve as the disconnecting means, *NEC 430.109(F)*. The attachment plug and receptacle must be horsepower rated. Most standard plugs and receptacles have been tested and given a horsepower-rating. This marking is not on the plug or receptacle so it is listed in Table 2-4. For motors of one-third horsepower or less, a horsepower rated plug and receptacle are not required. However, the ratings in Table 2-4 show that these plugs and receptacles are adequate.

If the motor is 2 hp or under and 300 volts or under, the disconnecting means can be a current-rated ac snap switch, seen in Figure 2-12, or a knife switch, seen in Figure 2-13, *NEC 430.109(C)*. The ac snap switch must be suitable for ac service only. Size according to the following criteria:

- **AC snap switch**: current rating greater than or equal to 1.25 times the motor full-load current rating.
- **Knife switch**: current rating greater than or equal to 2 times the motor full-load current rating.

Controller

A controller is a device used to start and stop a motor. Each motor must have its own controller except when there are several motors in one machine or system, *NEC 430.87*. The controller cannot be the same device used for the disconnecting means except for certain conditions as described later.

The controller need open only enough ungrounded wires to stop the motor—only one ungrounded wire needs to be opened for 120-volt or 240-volt single-phase service, *NEC 430.84*. Locate the controller within sight and within 50 feet of the motor. A disconnecting means must be in sight or within 50 feet of the controller, *NEC 430.102(A)*.

For ventilating fans with thermostats, variable-speed controllers, or timers, these items are considered to be the controller. The horsepower rating of these controllers must be greater than or equal to the motor's horsepower rating. Also, the controller's current rating must be greater than or equal to fan motor full-load current rating, as shown in Table 2-3. If not, use with a magnetic starter as shown in Figure 2-14.

EXAMPLE
2-1. Determining the required short-circuit protection for a 240-volt circuit.

Determine the short-circuit protection required for a 240-volt branch circuit with a 1/3-hp motor. The motor has no code letter.

SOLUTION

A 1/3-hp motor has a full-load current of 3.6 amps at 240 volts, Table 2-2. The maximum size breaker allowed for this motor circuit is 250% of the full-load current:

$$(2.50)(3.6 \text{ A}) = 9 \text{ amps}$$

This value is smaller than most readily available circuit breakers (which are typically 10 amps), so provide a time delay fuse at the motor disconnect to provide short-circuit protection for the motor circuit. Protect the conductors with a 15-amp circuit breaker in the SEP. The maximum size fuse is 175% of the full-load current:

$$(1.75)(3.6 \text{ A}) = 6.3 \text{ amps}$$

EXAMPLE
2-2. Determining the required short-circuit protection for a 120-volt circuit.

Determine the short-circuit protection required for a 1/3-hp motor on a 120-volt circuit. The motor has no code letter.

SOLUTION

A 1/3-hp motor has a full-load current of 7.2 amps at 120 volts, Table 2-2. The maximum size breaker allowed for the motor is 250% of the full-load current:

$$(2.50)(7.2 \text{ A}) = 18 \text{ amps}$$

A 15- or 20-amp (the next largest standard size can be selected) inverse time circuit breaker in the SEP would provide short-circuit protection. Another alternative would be to use a time delay fuse smaller than 175% of the full-load current at the motor disconnect:

$$(1.75)(7.2 \text{ A}) = 12.6 \text{ amps}$$

In addition, a 15-amp circuit breaker in the SEP is needed to provide short-circuit protection.

Table 2-4. Assigned horsepower ratings for attachment plugs and receptacles.

These values are for motors running at normal ratings. This material was extracted by MidWest Plan Service from Underwriters Laboratories Inc. *2008 General Information for Electrical Equipment.*

L-L: Motor connected line-to-line. L-N: Motor connected line-to-neutral.

Ampere rating	ac voltage rating	Phases	Poles	Horsepower rating	NEMA designation
15	125	1	2	1/2	1-15, L1-15, 5-15, L5-15
	250	1	2	2	2-15, 6-15, L6-15
	277	1	2	2	7-15, L7-15
	125/250	1	3	1-1/2	14-15
	250	3	3	2	11-15, L11-15, 15-15
	120/208	3	4	2	18-15
20	125	1	2	1	5-20, L5-20
	250	1	2	2*	2-20, L2-20, 6-20, L6-20
	277	1	2	2	7-20, L7-20
	480	1	2	3	L8-20
	125/250	1	3	2L-L*, 1L-N	10-20, L10-20, 14-20, L14-20
	250	3	3	3	11-20, L11-20, 15-20, L15-20
	480	3	3	3	L8-20
	120/208	3	4	2	18-20, L18-20, L21-20
	277/480	3	4	5	L19-20, L22-20
30	125	1	2	2	5-30, L5-30
	250	1	2	2*	2-30, 6-30, L6-30
	277	1	2	3	7-30, L7-30
	480	1	2	5	L8-30
	125/250 1	3	3	2L-L*, 2L-N	10-30, L10-30, 14-30, L14-30
	250	3	3	3	11-30, L11-30, 15-30, L15-30
	480	3	3	10	L12-30, L16-30
	120/208 3	3	4	3	18-30, L18-30, L21-30
	277/480 3	3	4	10	L19-30, L22-30
50	125	1	2	2	5-50
	250	1	2	3*	6-50
	277	1	2	5	7-50
	125/250	1	3	3L-L, 2L-N*	10-50, 14-50
	250	3	3	7-1/2	11-50, 15-50
	120/208	3	4	7-1/2	18-50
60	125/250	1	3	3L-L*, 2L-N	14-60
	250	3	3	10	15-60
	120/208	3	4	7-1/2	18-60

* Also suitable for 208-volt applications at indicated horsepower.

If the motor is portable and 1/3-hp or under, the receptacle and attachment plug can serve as the controller as well as the disconnecting means, *NEC 430.83(C)* and *430.109(C)*. A portable motor is one that is not fixed in place and is moved often. Size as described in the **Disconnecting Means** section.

For other motors, a motor starter works as the controller for most types of motor circuits. Obtain a motor starter with a horsepower rating greater than or equal to motor horsepower rating, *NEC 430.83*.

If the motor is 2 hp or less and 300 volts or less, an ac snap switch or knife switch can serve as the controller, *NEC 430.83*.

Figure 2-12. AC snap switch.

a. Manual

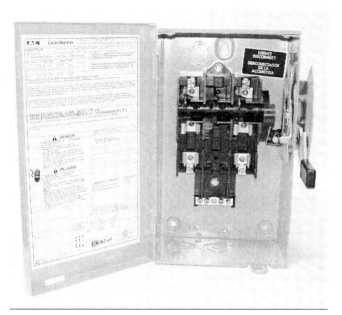

Figure 2-13. Knife switch with fuse holders.
Door is closed except during servicing.

b. Magnetic

Figure 2-14. Motor starters.
Typical motor starter with built-in overload devices.

The snap switch or knife switch should have the same rating as a disconnect.

A knife switch can be both the disconnecting means and controller, if it is operated by directly applying the hand to a lever, as detailed in Figure 2-13, *NEC 430.111*. The knife switch must be capable of opening all ungrounded (hot) wires and is protected by an overcurrent device. Size as described in the **Disconnecting Means** section.

Overload Protection

When an electric motor is starting or overloaded, it draws more current (amperes) than when delivering its rated horsepower. A motor will not be damaged by current larger than normal for a short time, but it will be burned out by current larger than normal after a longer period of time. Therefore, it is necessary to protect a motor with a device that will permit the high starting current for a short time, but will disconnect the motor if high current due to overload flows through the motor for a longer time. Most manually operated controllers

CHAPTER 2. BRANCH CIRCUITS

contain a heater device to trip the mechanism on overload. These devices do not function fast enough to protect against short-circuits, so the branch circuit must also be protected by circuit breakers or fuses as discussed earlier.

If the overload protection is a fuse, install one in each hot conductor for 120-volt or 240-volt single-phase service, *NEC 430.36*. For devices other than fuses, install it in one hot conductor for 120-volt or 240-volt single-phase service, *NEC 430.37*. If two are required, use a common trip.

For motors greater than 1 hp or automatic start (for example, fans on thermostats), either

- Obtain a motor with a built-in overload protection device, or
- Install an overload device separate from the motor. Size as follows:
    » *For motors marked for 40 C temperature rise*: Select an overload device with a current rating less than or equal to 1.25 times the motor nameplate full-load current rating.
    » *For motors with a service factor greater than or equal to 1.15*: Select an overload device with a current rating less than or equal to 1.25 times by the motor nameplate full-load current rating.
    » *All other motors*: Select an overload device with a current rating less than or equal to 1.15 times by the motor nameplate full-load current rating, *NEC 430.32*.

If the calculated value does not correspond to the standard sizes available, use the next larger size, but not greater than 1.4 times the motor full-load current rating for motors marked for 40 °C temperature rise or 1.3 times the motor full-load current rating for other motors, *NEC 430.32(C)*.

For motors less than or equal to 1 hp, manually started, within sight and within 50 feet of the controller, and permanently installed, the branch circuit breaker or time delay fuse can serve as the overload protection device as well as the short-circuit protection device, *NEC 430.32(D)*.

Branch Circuit Conductors

Conductors are wire or cable for conducting electric current. Wire usually refers to a single conductor, whereas cable refers to two or more conductors in the same sheathing. Copper is the most common and easy to work with material.

Aluminum branch circuit conductors require special wiring devices and techniques and are not discussed in this book.

Conductors carrying electric current are like pipes carrying water—the larger the pipe (wire), the more water (current) it can carry. The more current required by the end use, the larger the conductor must be. Wire is sized by the AWG system; the larger the conductor the smaller the gage number. Figure 2-15 illustrates various AWG wire sizes. For conductors larger than AWG 4/0, the size is in kcmils. The kcmil size is the actual cross-sectional area of the wire in thousands of circular mils.

Although the *NEC* allows AWG 14-CU conductor and 15-amp circuits for branch circuits, AWG 12-CU is highly recommended for agricultural-type buildings. Most **general-purpose** branch circuits can be AWG 12-CU. If the number of DCOs and/or lights does not exceed 10 and the average load of each DCO or light does not exceed 1.5 amps, AWG 12 is adequate for general-purpose branch circuits. For special purpose or dedicated individual circuits, AWG 14-CU may be adequate if voltage drop does not require a larger size.

However, check each **individual** branch circuit conductor size for the following criteria. Conductor size is based on two factors:

- Ampacity.
- Voltage drop.

Ampacity is the safe, current-carrying capacity of a conductor in amperes (A). Current flowing through a wire creates heat. If the amperage is too high, the wire may become hot enough to damage the insulation and start a fire.

Current in the circuit conductor creates heat which is wasted and also causes a voltage drop. For example, with 120 volts at the SEP, the voltage at a motor 100 feet away drawing 14 amps drops to about 115 volts if wired with AWG 12-CU wire. Voltage drop results in power or light loss and can result in inefficient appliance operation. Table 2-5 shows the effect of voltage drop on power and light loss.

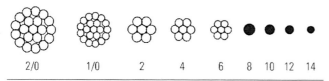

Figure 2-15. AWG wire sizes.
Diameters of typical wires without insulation. Note that the larger the gage number, the smaller the diameter.

Table 2-5. Effect of voltage drop on power and light loss.

Voltage drop, %	Light loss, %	Power loss, %
1	3	2.0
2	7	4.0
3	10	6.0
4	13	8.0
5	16	10.0
10	31	19.0
15	46	28.0

The larger the conductor, the less the voltage drop. Select branch circuit conductors large enough to limit voltage drop to 2%, as listed in Tables 2-6 and 2-7.

Use the following procedure to size a conductor:
- Determine the maximum amperage the conductor has to carry. Increase the maximum amperage by 25% if the branch circuit may carry sustained load for longer than 15 minutes. This includes all motor circuits.
- Determine the conductor length (one way). The one-way length is the distance from the SEP to the load via the conductor's route.
- Select the minimum conductor size to handle the amperage at a voltage drop not over 2%, based on the values listed in Tables 2-6 and 2-7.

Fundamental electrical terminology:
- **Voltage** is the pressure in the circuits (similar to water pressure in your plumbing system) and is measured in volts (V). Common voltages in buildings are 120 volts or 240 volts.
- **Amperage** is the current flowing through conductors and is measured in amps (A).
- **Wattage** is the power (work done over a period of time) and is measured in watts (W), or kilowatts 1 kW equals 1,000 watts.

Table 2-6. Minimum UF and NM cable copper conductor size for branch circuits.
Conductor length is one way. Use for individual circuits and for general-purpose circuits that do not meet the criteria in Figure 2-3. Based on the conductor ampacity or 2% voltage drop, whichever is limiting.

Load (amps)	Conductor size (AWG) for 120-volt service									
	Conductor length (feet)									
	30	40	50	60	75	100	125	150	175	200
5	12	12	12	12	12	12	12	10	10	10
7	12	12	12	12	12	12	10	10	8	8
10	12	12	12	12	10	10	8	8	8	6
15	12	12	10	10	10	8	6	6	6	4
20	12	10	10	8	8	6	6	4	4	4
25	10	10	8	8	6	6	4	4	4	3
30	10	8	8	8	6	4	4	4	3	2
35	8	8	8	6	6	4	4	3	2	2
40	8	8	6	6	4	4	3	2	2	1
45	6	6	6	6	4	4	3	2	1	1
50	6	6	6	4	4	3	2	1	1	1/0
60	4	4	4	4	4	2	1	1	1/0	2/0
70	4	4	4	4	3	2	1	1/0	2/0	2/0
80	3	3	3	3	2	1	1/0	2/0	2/0	3/0
90	2	2	2	2	2	1	1/0	2/0	3/0	3/0
100	1	1	1	1	1	1/0	2/0	3/0	3/0	4/0

Load (amps)	Conductor size (AWG) for 240-volt service									
	Conductor length (feet)									
	50	60	75	100	125	150	175	200	225	250
5	12	12	12	12	12	12	12	12	12	12
7	12	12	12	12	12	12	12	12	10	10
10	12	12	12	12	12	10	10	10	10	8
15	12	12	12	10	10	10	8	8	8	6
20	12	12	10	10	8	8	8	6	6	6
25	10	10	10	8	8	6	6	6	6	4
30	10	10	10	8	6	6	6	4	4	4
35	8	8	8	8	6	6	4	4	4	4
40	8	8	8	6	6	4	4	4	4	3
45	6	6	6	6	6	4	4	4	3	3
50	6	6	6	6	4	4	4	3	3	2
60	4	4	4	4	4	4	3	2	2	1
70	4	4	4	4	4	3	2	2	1	1
80	3	3	3	3	3	2	2	1	1	1/0
90	2	2	2	2	2	2	1	1	1/0	1/0
100	1	1	1	1	1	1	1	1/0	1/0	2/0

Table 2-7. Minimum THW, THWN, RHW and XHHW copper conductor size for branch circuits.
Conductor length is one way. Use for individual circuits and for general-purpose circuits that do not meet the criteria in Figure 2-3. Based on conductor ampacity or 2% voltage drop, whichever is limiting.

Load (amps)	Conductor size (AWG) for 120-volt supply — Conductor length (feet)									
	30	40	50	60	75	100	125	150	175	200
5	12	12	12	12	12	12	12	10	10	8
7	12	12	12	12	12	10	10	8	8	8
10	12	12	12	12	10	10	8	8	6	6
15	12	12	10	10	10	8	6	6	4	4
20	12	10	10	8	8	6	6	4	4	4
25	10	8	8	8	6	6	4	4	4	3
30	10	8	8	6	6	4	4	3	2	2
35	8	8	8	6	6	4	3	2	2	1
40	8	6	6	6	4	4	3	2	1	1
45	8	6	6	4	4	3	2	1	1	1/0
50	8	6	6	4	4	3	2	1	1/0	1/0
60	6	6	4	4	4	2	1	1/0	2/0	1/0
70	4	4	4	3	2	1	1/0	2/0	2/0	3/0
80	4	4	4	3	2	1	1/0	2/0	3/0	3/0
90	3	3	3	2	1	1/0	2/0	3/0	2/0	4/0
100	3	3	3	2	1	1/0	2/0	3/0	4/0	4/0

Load (amps)	Conductor size (AWG) for 240-volt supply — Conductor length (feet)									
	50	60	75	100	125	150	175	200	225	250
5	12	12	12	12	12	12	12	12	12	12
7	12	12	12	12	12	12	12	10	10	10
10	12	12	12	12	12	10	10	8	8	8
15	12	12	12	10	10	8	8	8	6	6
20	12	12	10	8	8	8	6	6	4	4
25	10	10	10	8	8	6	6	6	4	4
30	10	10	8	8	6	6	4	4	4	4
35	8	8	8	6	6	4	4	4	4	3
40	8	8	8	6	6	4	4	4	3	3
45	8	8	6	6	4	4	4	3	2	2
50	8	8	6	6	4	4	3	3	2	2
60	6	6	6	4	4	3	2	2	1	1
70	4	4	4	4	3	2	2	1	1	1/0
80	4	4	4	4	3	2	1	1	1/0	1/0
90	3	3	3	3	2	1	1	1/0	1/0	2/0
100	3	3	3	3	2	1	1/0	1/0	2/0	2/0

To determine current flow for electrical devices, such as incandescent lights and electric resistance heaters, use the following procedure:

Equation 2-1. Determining current flow.

$$I = \frac{P}{V}$$

Where:
- I = current, amps
- P = power, watts
- V = voltage, volts

For example, a 600-watt heater on a 120-volt circuit uses

$$I = \frac{600 \text{ W}}{120 \text{ V}} = 5 \text{ amps}$$

Guidelines for computing the amperage in each branch circuit are as follows:

- **Lights:** Allow 1.5 amps per light fixture for 120-volt service. For heat or flood lamps larger than 180 watts, use the actual figure. For example, a 250-watt heat lamp on 120-volt service consumes:

$$I = \frac{250 \text{ W}}{120 \text{ V}} = 2.1 \text{ amps}$$

- **Outlets:** Allow 1.5 amps per DCO on 120-volt service unless the outlet will be used for motors, heat lamps, or other large power users. Use the actual value of the load for SPOs that supply large power users.

- **Motors:** For a branch circuit with only one motor, size the conductor for 125% of the motor's full-load current rating, *NEC 430.22(A)*. For example, for one, 6-amp motor on a circuit:

$$(1.25)(6A) = 7.5 \text{ amps}$$

For more than one motor on a circuit, rate the largest motor at 125% and add the other motors at 100% of their full-load current rating, *NEC 430.24*. Use current ratings in Table 2-3, not nameplate ratings. If other loads, such as lights or heaters, are on this circuit, add 100% of their full-load current to the above calculations, *NEC 430.24)*.

CHAPTER 3

Service Entrance

THE SERVICE ENTRANCE of a building consists of the following elements:
- Service entrance panel (SEP).
- Main breaker.
- Circuit breakers.
- Ground connection.
- Service entrance conductors.

Circuit Breakers

Most SEPs contain one large, two-pole main breaker that protects the entire installation and disconnects it from the power source. However, there must also be breakers to protect the individual branch circuits—single-pole breakers for 120-volt and two-pole for 240-volt circuits. The branch circuit breakers are purchased separately from the SEP and plugged into the SEP as needed.

Sizing Circuit Breakers

Circuit breakers are rated in amperes. Except for motor circuits, the circuit breaker must have a rating in amperes **not** greater than the allowable current-carrying capacity (ampacity) of the conductors it protects. Table 3-1 lists the ampacity of copper conductors. For example, AWG 12 CU wire has an ampacity of 20 amps, so use a circuit breaker with an amperage rating of 20 amps or less. Also, the voltage rating of the circuit breaker must be greater than or equal to the circuit voltage.

Installing Circuit Breakers

For 120-volt circuits, attach the black hot wire from the branch circuit cable to the terminal on the circuit breaker, and attach the circuit neutral to the neutral bar and the equipment grounding conductor to the grounding bar. For

Table 3-1. Ampacity of copper conductors (amps).
Based on NEC Table 310.15(B)16 (formerly 310.16) and Article 240.4(D).

Wire size (AWG-CU)	THWN, RHW, UF NM cable	XHHW wire, and USE cable*
14	15	15
12	20	20
10	30	30
8	40	50
6	55	65
4	70	85
3	85	100
2	95	115
1	110	130
1/0	125	150
2/0	145	175
3/0	165	200
4/0	195	230

*AWG 14 CU wire is recommended only for specific equipment—not for general purpose branch circuits.

240-volt circuits, attach one hot wire to each circuit breaker terminal. Attach the equipment grounding conductor to the grounding bar, as illustrated in Figure 3-1.

Ground Fault Circuit Interrupters

There are special circuit breakers and receptacles with ground fault interruption (GFCI). Figure 3-2 shows a typical GFCI circuit breaker. These devices protect humans and animals from dangerous shock by opening the circuit almost immediately upon detecting a ground fault (current is escaping to ground somewhere in the circuit). A ground fault usually results from a breakdown in the insulation of an electrical device. The GFCI monitors the current going

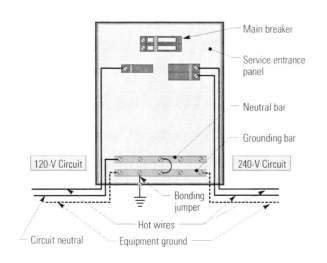

Figure 3-1. Wiring of branch circuit breakers in SEP.

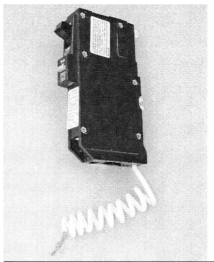

For 120-volt circuits, both the circuit neutral and hot wire connect to the GFCI.

The GFCI has a white lead to connect to the neutral bar.

Figure 3-2. GFCI circuit breaker.

out and the current returning. If these current levels are not equal, then a ground fault is detected, and the hot wire is disconnected in about 0.125 seconds.

The *NEC* requires that GFCIs be used in damp or wet areas of agricultural-type buildings *NEC 547.5(G)*. GFCIs are also required for outlets located outside, in toilet facilities, wash areas, and shops where floors may be wet; GFCIs can be used for any circuit. The GFCI protection can be provided by a GFCI circuit breaker or by a GFCI receptacle. Some current leaks from all circuits and may be detected by the GFCI causing it to trip. Such nuisance tripping is most common when extension cords are used to effectively lengthen the circuit. GFCIs are recommended for old power washers and are required for new power washers, *NEC 422.49*.

Receptacle GFCIs can be used to replace 120-volt receptacles which have no equipment ground. In these applications, the circuit can be used without an equipment ground, *NEC 406.4 (D)*.

Service Entrance Panel (SEP)

If possible, avoid locating the SEP in humid, dusty, and/or corrosive environments. If the building has an entry, office, or utility room with none of the above environments, locate the SEP there.

If the SEP must be placed in the same room as the animals, use a weathertight or raintight plastic cabinet with a National Electrical Manufacturers Association (NEMA) 4X rating. In feed handling areas, use dustproof boxes (NEMA 4X). Surface mount the SEP with about a 1/4-inch gap between the mounting panel and the wall for ventilation and sanitation, *NEC 300.6(D)*. Install all branch circuit cables through knockout holes at or near the bottom of the SEP. This arrangement, as shown in Figure 3-3, tends to keep condensed water from running down the cable or in the conduit and into the SEP.

Figure 3-4 shows the interior of an SEP that includes the main breaker, neutral bar, grounding bar, and a circuit breaker panel. The SEP manufacturer provides installation instructions. The SEPs are made in standard sizes of 100, 125, 150, 200, 225, 300, 400, and 600 amps. Run all the branch circuit conductors into the SEP before doing the final wiring in the SEP. Install conductors in the SEP in a neat and orderly manner. Excess lengths of wire result in undue heat buildup in the SEP.

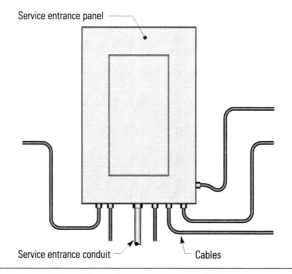

Figure 3-3. All cable and conduit enter SEP at or near the bottom.

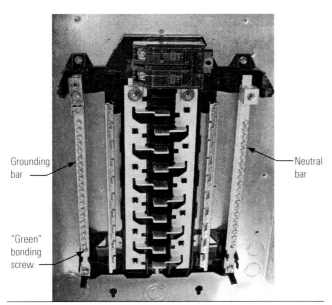

Figure 3-4. Service entrance panel (SEP) box.

Sub Panels

Many installations need a sub panel in another room from the main SEP. Figure 3-5 illustrates a sub panel in the same building as the main panel and also shows a sub panel in a building other than the one in which the main panel is located. The sub panel is essentially handled as a 240-volt branch circuit. Connect the sub panel feeder wires to a two-pole, 240-volt circuit breaker in the main panel. The sub panel does not need a main breaker if the feeder conductors are protected by a circuit breaker in the main panel, and if the amperage capacity of the circuit breaker protecting the feeder conductors does not exceed the amperage capacity of the sub panel, *NEC 408.36*.

However, a main breaker at the sub panel is desirable, and if all the above conditions are not met, a main breaker is required. Size the sub panel main breaker to have sufficient capacity for all the branch circuits to be connected to it. Use the same methods described for sizing the main SEP. The sub panel feeder cable capacity, listed in Table 3-4, should at least match the sub panel amperage capacity. Refer to the panel rating and manufacturer's instructions.

For grounding, run the circuit neutral and equipment grounding conductor to separate bars in the sub panel. If the sub panel is in a separate building, as shown in Figure 3-5, ground it as shown. The neutral bar and grounding bar are not bonded together in a sub panel. When the sub panel is in another building, there must be a main breaker (disconnecting means) inside or outside of the building where the conductors pass through the building wall, *NEC 225.32*.

Main Breaker

Main breakers are rated at 30, 40, 50, 60, 70, 100, 125, 150, 175, and 200 amps and above. Minimum size for agricultural-type buildings is 60 amps, *NEC 230.79(D)*. Most modern agricultural-type buildings require at least 100-amp services. To determine what size main breaker to use, estimate the building's maximum overall power usage. Include any known future loads—also consider increasing the main breaker size to allow for any future expansion.

Use the following to help estimate the amperage load of various electrical equipment:

- DCOs—use 1.5 amps per DCO at 120 volts. This accounts for all plug-in appliances such as power tools and portable fans.
- Motors—use a value equal to the full load current of the largest motor times 1.25; plus 100% of the full load current of all the other motors. If you have two or more motors of equal size on the same circuit, apply the 125% factor to only one. If the motors are rated in horsepower, use Table 2-3 to estimate full load current in amperes.
- All other permanently connected equipment—use 100% of the rated full load current. If an electric resistance heater is rated in watts instead of amps, divide the wattage value by the service voltage (120 volts or 240 volts) to get the amperage.

Make a list of all electrical equipment and its amperage load. The SEP is wired to 240-volt service. Use the following equation to convert the amperages of all electrical equipment to 240-volt service:

Equation 3-1. Converting amperage from 120 volts to 240 volts.

$$\text{Amperage at 240 V} = \frac{\text{Amperage at 120V}}{2}$$

Generally, the **maximum** current load on the main breaker will not be the sum of all the electrical equipment amperages because not all of the electrical equipment will be on at the same time. You will need to determine which equipment could conceivably be on at the same time and

a. Sub panel in same building as main panel
Main breaker not required in the sub panel if the feeder cables are protected by a circuit breaker in the main panel and if the amperage capacity of that circuit breaker does not exceed the amperage capacity of the sub panel.

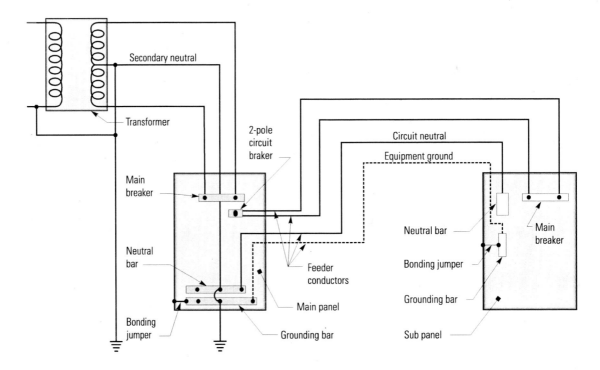

b. Sub panel in another building

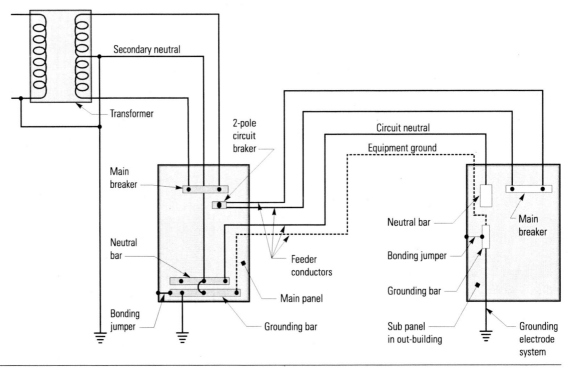

Figure 3-5. Wiring sub panels.
If the feeder conductor cable is run underground, the equipment ground must be an insulated copper conductor. Install the feeder cable with extreme care, because there is no way to monitor the effectiveness of the equipment ground after it is in place.

which equipment will be off when other related equipment is on. This requires good judgment and knowledge of the building operation. For example, in a swine building it is conceivable that all the lights and DCOs would draw power at the same time. However, the heater and the warm weather fans would not run at the same time.

If the total current load for the building is greater than 60 amps, you may be able to reduce the size of the main breaker to allow for the fact that all loads will not be on at the same time. If the total current load for the building is less than 60 amps, you must size the main breaker at least as large as the sum of all the amperage loads in the building. Agricultural-type buildings require at least a 60-amp service entrance.

All the loads that are likely to operate at the same time are defined as operating **without diversity**. The loads without diversity and all other loads are multiplied by a demand factor, which is listed in Table 3-2, and summed to get the overall load on the service entrance. Example 3-1 illustrates this procedure.

Table 3-2. Demand factor.
Based on *NEC Table 220.102*.

Ampere load at 240 V	Demand factor (%)
Loads expected to operate without diversity, but not less than 125% full-load current of the largest motor and not less than the first 60 A of load.	100
Next 60 A of all other loads	50
Remainder of other loads	25

EXAMPLE
3-1. Determining the main breaker size for a farrowing building.
Determine the main breaker size for a swine farrowing room with the following electrical loads:

SOLUTION

Electrical equipment	Amperage at 240 V
Lights: (24 lights)(1.5 A per light) = 36 A at 120 V	18.0 A
Heat lamps: $(10 \text{ DCO})\left(2 \text{ lamps}/\text{DCO}\right)\left(\dfrac{250 \text{ W}/\text{lamp}}{120 \text{ V}}\right) = 41.6$ amps	20.8 A
Outlets: (6 DCOs)(1.5 A per DCO) = 9.0 amps	4.5 A
Heaters: $(2 \text{ heaters})\left(\dfrac{3{,}000 \text{ W}/\text{heater}}{240 \text{ V}}\right) = 25.0$ amps	25.0 A
Fans, cold weather: (Table 2-4, 1/3 hp fans): (2 fans)(3.6 A per fan) = 7.2 amps	7.2 A
Fans, warm weather: (Table 2-4, 1/2 hp fans): (3 fans)(4.9 A per fan) = 14.7 amps	14.7 A
Auger motors: (Table 2-4, 3/4 hp motors): (1.25*)(6.9 A) + 6.9 A = 15.5 A	15.5 A
Total amperage:	**105.7 A**

*Multiply the table value of one of these motors times 1.25, as these are the largest motors in the building.

Continued

Equipment operating without diversity at 240 V:	
Lights	18.0 A
Heat lamps	20.8 A
Outlets	4.5 A
Heaters	25.0 A
Fans, cold weather	7.2 A
Auger motors	15.5 A
Total load without diversity	**91.0 A**

It is conceivable that all the lights and DCOs, the auger motors and the cold weather fans could operate at the same time, that is, without diversity. The heaters and warm weather fans would not operate at the same time. You must use 100% of the loads without diversity, Table 3-2.

Use a service entrance main breaker rated greater than 91.0 amps; 100 amps is the next larger size. A 200-amp main breaker may be a better option because it will allow for future expansion without adding much additional cost. Loads without diversity exceeding 80 amps must use a main breaker greater than 100 amps. Loads without diversity would be multiplied by 1.25 to account for continuous load.

Service Entrance Wiring Diagram

After planning the branch circuits, draw a wiring diagram of the SEP. The SEP diagram illustrates panel capacity, main breaker capacity, circuit breaker capacities and locations, and a schematic drawing of the wiring within the SEP for an even distribution of loads. Evenly distribute the 120-V loads between each of the hot service entrance wires, that is, distribute the circuit breakers so there are about the same number on each side of the SEP. A double breaker that occupies the same space as two single breakers is required for 240-V service. See **Chapter 8, Design Examples**, for example SEP wiring diagrams.

Proper tightening of screws to make electrical connections is very important. Most SEPs have torque specifications for all screw terminals in the SEP.

Ground at the Service Entrance

At the SEP, the equipment grounds (grounding conductors) are connected to the grounding bar. The grounding bar, the circuit neutrals (120-volt service), and the neutral of the three incoming service entrance conductors are all connected to the neutral bar. The neutral bar is connected (grounded) to the grounding electrode system by the grounding electrode conductor. The power supplier also grounds the neutral of the service conductors at the power pole. Figure 3-6 gives an example of a typical electrical grounding system.

The neutral bar has large connectors for the larger size incoming neutral and the grounding electrode conductor, plus many smaller connections for the circuit neutral of each 120-volt circuit. Ground the SEP cabinet to the neutral bar, as shown in Figure 3-7. This connection is often provided in the SEP by a green colored screw which must be tightened or by a ground strap.

Ground all conductive equipment (stanchions, partitions, pipes, etc) in the building to the ground bar with an equipment grounding conductor. Otherwise, an electrical fault to

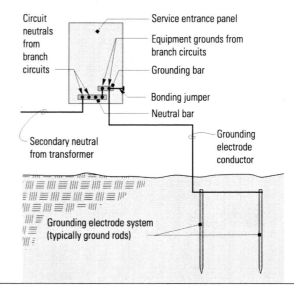

Figure 3-6. Electrical grounding system.

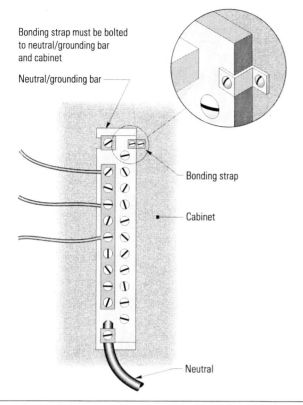

Figure 3-7. Grounding the service entrance panel to the neutral/grounding bar with a bonding strap.
A bonding strap is also called the bonding jumper. Run all equipment grounds to the grounding bar.

Note: A neutral/grounding bar is not a recommended arrangement for rural buildings. Instead use separate neutral and grounding bars.

EXAMPLE
3-2. Determining the main breaker size for a dairy barn.
Determine the main breaker size for a dairy barn with the following electrical loads:

Equipment operating without diversity at 240 V:	
Some lights and DCOs	10.0 A
All fans	26.8 A
Milk cooler	10.4 A
Water heater	12.5 A
Fans, cold weather	7.2 A
Silo unloader	35.0 A
Total load without diversity	**94.7 A**

Demand load at 240 V:	
Load without diversity at 100%	94.7 A
Next 60 amps at 50%	
(0.5) (60 A)	30.0 A
Remaining load at 25%	
(0.25) (185.1 – 94.7 A – 60 A)	7.6 A
Total load	**132.3 A**

SOLUTION
It is conceivable that some of the lights and DCOs, all the fans, the milk cooler, the water heater, and the silo unloader may operate at the same time, that is, without diversity. The barn cleaner, water pump, and tank truck SPO would probably not operate at the same time as the milker and milk pump.

Use a service entrance main breaker with at least a 150-amp rating. Using a 200-amp rated main breaker would allow for any future expansion without adding much to costs.

ungrounded equipment could easily be fatal. If there is metal, underground water piping, connect the grounding electrode conductor to the water pipe as well as to the ground rod (electrode). This will reduce the chance of stray voltage in the water piping system.

The electrical system must have its own grounding electrode system separate from the grounding electrode of a lightning protection system, *NEC 250.60*. However, it is required that the electrical and lightning grounding electrodes be bonded together, *NEC 250.106*.

Grounding Electrode Conductor
The grounding electrode conductor must be corrosion resistant—copper wire is preferred because it is less likely to corrode under agricultural conditions. Make the grounding electrode conductor one continuous length without splices, *NEC250.64(C)*. Splices are permitted in some instances. Grounding electrode conductor sizes are based on the size of the largest service entrance conductor, listed in Table 3-3. For example, a building with a 175-amp service entrance

CHAPTER 3. SERVICE ENTRANCE

Table 3-3. Sizing grounding electrode conductors.

Size of largest service entrance conductor		Recommended grounding electrode conductor size*
Copper	Aluminum	
AWG 1/0 or smaller	AWG 3/0 or smaller	AWG 6-CU
AWG 2/0 through AWG 3/0	AWG 4/0 through 250 kcmil	AWG 4-CU
Over AWG 3/0 through 350 kcmil	Over 250 kcmil through 500 kcmil	AWG 2-CU

*If a ground rod is used for the grounding electrode, the *NEC* states that the grounding electrode conductor need not be larger than AWG 6-CU. *NEC 250.66A*. If the service entrance is large enough to require a larger grounding electrode size, consider installing a grounding electrode system *NEC 250.53*.

that has AWG 2/0 CU service entrance conductors requires an AWG 4 CU grounding electrode conductor. A building with a 250-amp service entrance and 350-kcmil AL (aluminum) service entrance conductors requires an AWG 2 CU grounding electrode conductor.

If the grounding electrode conductor is not subject to physical abuse, it can be bare; fasten it rigidly to the building. If the grounding electrode conductor may be exposed to abuse, enclose it in a nonmetallic rigid conduit. The conduit must meet the criteria given in the nonmetallic (plastic) conduit section. Prevent animals from chewing on conduit, *NEC 250.64(B)*.

Grounding Electrode System

If a metal underground water pipe extends at least 10 feet into earth, use it plus an additional electrode for the grounding electrode system. See the last paragraph of this section and Figure 3-8a for a new additional recommendation for a grounding electrode system. Additional electrodes can be the metal building frame if it makes good contact with earth, 20 feet of bare AWG 4 CU wire encased in concrete (such as in the foundation), or a loop of AWG 2 CU bare wire buried in the ground, *NEC 250.52*.

If the above grounding electrodes are not available, use a dedicated ground rod with a steel core (Figure 3-8b) and a copper layer on the outside, driven at least 8 feet into the ground. It must be at least 1/2 inch in diameter. Galvanized water pipe (at least 8 feet long and at least 3/4 inches in diameter) is acceptable according to the *NEC*, but not recommended. If the ground rod is copper coated, use a copper or brass clamp. If it is galvanized pipe, use a galvanized iron clamp. Clamps must be rated for burial.

Use extreme care when installing grounds so the grounding system is well protected from future damage. Locate the ground rod near the service entrance so the grounding electrode conductor is as short and direct as possible. Dig a 1-foot deep trench from the building to the ground rod location. Locate the rod about 2 feet from the building in a place where the earth will remain wet. A good location is where rain from a roof accumulates. Drive the ground rod at least 8 feet into the ground.

Drive the rod in so the top of the rod is only a few inches above the bottom of the trench. The ground wire runs down the side of the building to the bottom of the trench, then to the clamp on the rod, as illustrated in Figure 3-8c. After making the connections, fill the trench so the rod, clamp, and bottom end of the grounding electrode conductor are buried. Seepage from manure must not saturate the ground around the rod, because it can severely corrode the ground connection.

When several rods are used, keep them at least 16 feet apart. The 6-foot minimum spacing in the *NEC* is for personal safety only. The 16-foot spacing maximizes current dissipation of the ground rods. Connect all of the rods together with an unspliced grounding electrode conductor, *NEC 250.64(C)*.

If a reinforced concrete foundation is planned, make a connection available from that reinforcing rod to obtain an excellent low-resistance electrode. Metal building frames, now in common use, also make good ground electrodes when they are adequately anchored to the concrete foundation.

If the building will have electronic equipment, it will be wise to protect the electrical system from lightning surges. This can be accomplished by adding a ground plate to the grounding electrode system. See Figure 3-8a. A ground plate (250.53(H)) has a very large surface area compared to a ground rod. Since lightning is very high frequency and travels on the skin of the electrode, this type of installation has proven to be very effective in reducing lightning damage to electronic (computer-type) installations.

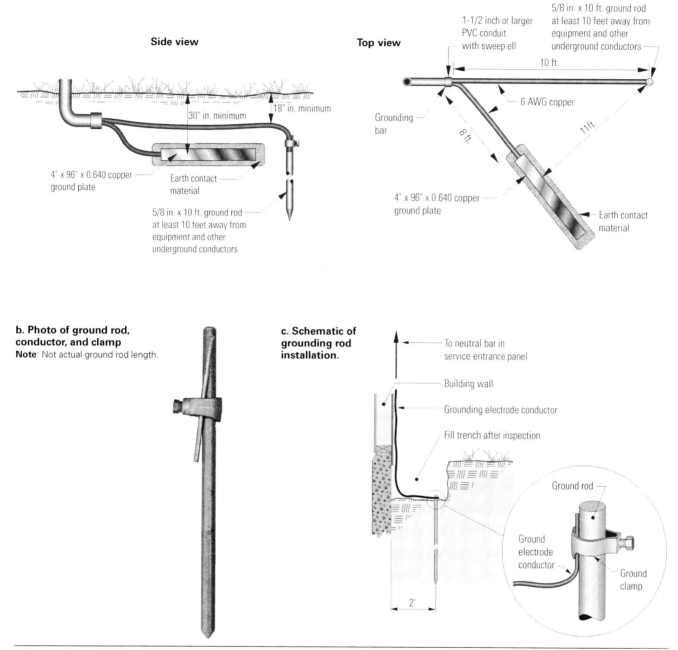

Figure 3-8. Ground rod installation.

Service Conductors

Power from the electric utility is delivered to the transformer and main meter on the site. Figure 3-9 depicts a typical power route from plant to load. The main meter is often on a power pole, but may be from a pad-mounted transformer. Service drop conductors carry power overhead from the power pole to the service drop on the building. Service lateral conductors carry power underground from the power pole to the service entrance conductors. Service entrance conductors carry power from the service drop to the SEP.

Size the service conductors to carry the amperage capacity of the main breaker, as listed in Table 3-4. Four-wire (240-volt) or three-wire (120-volt) service is strongly recommended for most agricultural-type buildings. These services

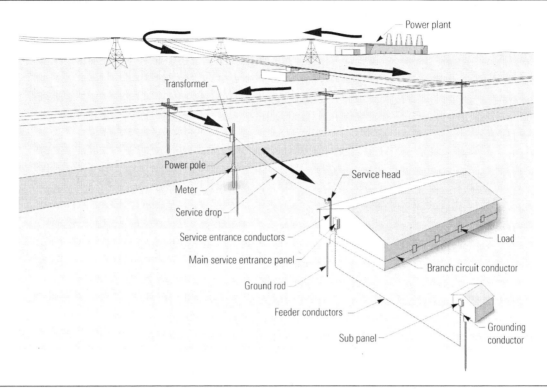

Figure 3-9. Power route from plant to load.

Table 3-4. Amperage capacity of service entrance conductors.
For 240-volt service and type **RHW**, **THWN**, and **XHHW** wires in conduit. Based on *NEC Table 310.15 (B)(16)*.

Panel service size (amps)	Conductor size	
	Copper	Aluminum
65	AWG 6	AWG 4
100	AWG 3	AWG 1
150	AWG 1/0	AWG 3/0
175	AWG 2/0	AWG 4/0
200	AWG 3/0	250 kcmil
250	250 kcmil	350 kcmil
300	350 kcmil	500 kcmil

provide a grounding conductor as well as a neutral for safety. The hot wires are typically both black, or one black and one red. If the building power requirements are high and three-phase power is available, it should be used.

Service entrance conductors run through conduit from the service drop to the SEP. The conduit enters at or near the bottom of the SEP. The conduit is sealed at both ends with sealing compound to fill all air voids between wires and the sides of the conduit to prevent moist air from getting into the conduit where the moisture could condense. It is strongly recommended that the service drop conductors or service lateral conductors, service entrance conductors, and SEP be installed only by a qualified electrician.

Power to Building

As shown in Figure 3-9, electricity is generated at the power plant and transmitted over utility high voltage power lines to the site's transformer. The transformer reduces the high voltage to the 120-volt and 240-volt service required for most operations. Sites supplied with three-phase power may have only 208-volt service instead of 240 volts. Wires from the power supplier's transformer end at a pole in the yard or at a pad-mounted transformer. Service drop or service lateral conductors carry current from the power pole or pad-mounted transformer (with main meter) to various points around the site.

Pole- or pad-mount metering provides a central distribution point for electric service to the site. It reduces the length of wire required to serve multiple buildings; it simplifies wiring repairs and additions, and it allows smaller feeder conductors to each building. Services to buildings can be underground or overhead. Consider locating the metering based on the following conditions:

- A large load is planned for two or more buildings in addition to the site's residence.
- Buildings are widely separated.

The conductors from the power pole or pad-mounted transformer are the customer's responsibility. It is important to consider power loss in the service drop or service lateral conductors. Design these conductors for 2% voltage drop. The general recommendation is to **limit voltage drop from the meter to the farthest loads to 5%**. If there is 2% loss in the service conductors, 2% loss in the feeder conductors and 2% loss in the branch conductors, then the 5% goal has not been achieved. However, a 6% voltage drop may be the best the electrical designer can expect.

Table 3-5 lists the recommended size of aluminum conductors for service drops or service laterals to limit the voltage drop to 2%. Use the one-way distance to the building; the table incorporates twice the distance. Note that the conductor size becomes very large with distance. When the required conductor size becomes more than 500 kcmil, it will require paralleling smaller conductors or moving the service closer to that building.

Values in Table 3-5 were calculated using Equation 3-2:

Equation 3-2. Conductor cross-sectional area.

$$CM = c \frac{25IL}{V_d}$$

Where:
- CM = Conductor cross-sectional area (circular mils)
- C = Constant (14.0 for copper; 1.6 for aluminum)
- I = Single phase current (amps)
- L = Length (one-way) of service Drop or service lateral (feet)
- Vd = Desired voltage drop (volts)

The conductor size is selected from a conversion table listing circular mills and AWG wire size. The final decision is based on two factors. The conductor must first be large enough to meet the electrical code minimums, and then the size can be increased to meet the voltage drop design when the distance calculation results in a circular mill larger than the code minimum.

Underground wiring is recommended for distribution service from the meter location to buildings. Underground wiring is simple to install with trenching equipment and special underground cable. Underground wiring allows tall machinery to move freely around the site without hazard of striking overhead wires. It also reduces the risk of service interruptions by ice storms, falling branches, or high winds. However, underground wiring is subject to damage from burrowing animals, rocks, corrosion, and trenchers, and it is more difficult to service or replace. To prevent accidental digging into buried wires, make a drawing of the site showing wire locations and keep it updated. Consider installing the underground wires in rigid, nonmetallic conduit.

Underground Service Entrance (USE) cable is the most common cable type for running buried feeder circuits from meter to building. Have all service drop conductors installed by a qualified electrician. As the number of buildings on your site increases, check to make sure the capabilities of your main service are not exceeded. Consult with your local power supplier.

Table 3-5. Aluminum wire size for a 2% voltage drop.
Values based on Equation 3-2.

Load (amps)	Distance to load (feet)									
	50	75	100	150	200	250	300	400	500	800
30	8	6	6	4	3	2	1	1/0	2/0	4/0
60	3	3	3	1	1/0	2/0	3/0	4/0	250	400
75	2	2	2	1/0	2/0	3/0	4/0	250	400	500
100	1	1	1	2/0	3/0	4/0	250	400	500	700
150	3/0	3/0	3/0	4/0	250	400	400	500	700	1000
200	250	250	250	250	400	500	500	700	900	—
250	350	350	350	400	500	700	800	1000	—	—
300	500	500	500	500	500	700	800	1000	—	—
400	900	900	900	900	900	900	1000	—	—	—

CHAPTER 4
Standby Power

STANDBY GENERATORS reduce the risk of loss and inconvenience associated with power outages. A double-throw transfer switch is required at the main farm service entrance, or at the building where standby service is needed. This transfer switch assures that the standby power source is always isolated from incoming power lines. This prevents generator damage when utility power is restored and keeps generated power from feeding back over the supply lines when a line worker expects it to be safe.

Most livestock producers should have standby generators as insurance against power failure. On many farms, the risks are large. For example, animals in enclosed buildings need a continuous supply of fresh air; animals need to be fed and watered regularly, and most heating equipment requires electricity to operate. In addition, many electric suppliers will provide energy at reduced rates if a generator can be run in place of utility electrical service during peak power use periods. This arrangement is practical only for permanently installed generators.

Develop instructions or procedures for the standby power system. Post a copy of the instructions near the generator. This provides a checklist for the regular operators and a procedure to follow in case no regular operators are available.

Generator Types

Figure 4-1 depicts the two basic types of standby generators:

1. **Engine-driven generators** are usually permanently installed, *NEC 702*, and can be made to start automatically when a power outage occurs.
2. **Tractor-driven generators** are usually portable, *NEC 250.34*, but can be permanently installed. Power is supplied through the power-take-off (PTO) shaft of a tractor.

a. Engine driven

b. Tractor driven

Figure 4-1. Generator types.

Engine-driven generators are more convenient because they are in place when needed. Automatic start units offer more protection because no one has to be around to detect the power loss and start the generator. Their main disadvantage is that they cost two to three times as much as tractor-driven generators. Also, the generator must be started (exercised) at least every month to ensure that it will start when needed.

Tractor-driven generators are inconvenient to position and hook up, especially in adverse weather when they are usually needed. The tractor may not start in cold weather.

Most generators provide 120/240-volt, three-wire, single-phase electrical service. If your farm has three-phase service, consult with your power supplier about the type of generator required. Purchase a generator capable of regulating the voltage so the voltage does not drop more than 10% between no load and full load. Otherwise, electric motors may overheat or operate improperly.

Generator Sizing

Determine if the generator is for a full-load system or a part-load system. A full-load system starts and operates all loads on the system. Automatic start generators must be sized for a full-load system. Part-load systems power only the critical loads on the farm (for example, minimum ventilating fans, milk cooling, etc.). Someone must be around to detect power failure, shut off all the loads, start the generator, and sequentially start the required loads. This can be done automatically, but is more expensive.

Generators are usually rated in watts. Two ratings are often listed:
1. A **continuous rating** for normal operation.
2. A **higher maximum rating** to allow for power surges created by motors that start and stop intermittently.

The maximum rating is only for short-term loads, but the actual meaning of the maximum rating varies with manufacturer.

For **full-load systems:**
1. List all the loads and their wattage; Table 4-1 lists the wattages of typical farm equipment. Use the starting wattage for motors listed in Table 4-2. Add all the wattages together. If there is one group of loads that does not run at the same time as another group, do not include both groups in the total wattage. Use the group with the largest wattage.
2. Add capacity for future loads. Standby generators are a long-term investment. Evaluate your inventory of electrical equipment and consider adding items you might acquire in the next 5 to 10 years.
3. Select a generator with a continuous rating that is at least as large as the total wattage. If there are motors in the system, you may be able to select a generator based on its maximum rating, because motors consume the higher starting wattage for only a short time. Check with the generator manufacturer.

For **part-load systems:**
1. Determine which loads are critical (for example, minimum ventilation, milking, water pumping, lighting, milk cooling, heating, etc.).
2. List all the critical loads and indicate their wattage, shown in Table 4-1. Include both the starting and running wattage for motors, as listed in Table 4-2.
3. Determine the load starting sequence and list in steps. The starting sequence may depend on the function of each load. Try to start the largest motor first and work down to the smallest motor; then start the other loads. Motors require four to six times as much power to start as to run, so the generator can be much smaller if the motors are started sequentially rather than all at once.
4. For each step, add the running wattage of items already operating to the starting wattage of the item being started in that step. Choose the largest wattage value out of all the steps for sizing the generator. If any of the motors stop and start intermittently, you may have to use their starting wattage in all the steps depending on the generator's short-term surge capabilities.
5. Add capacity for future loads.
6. Compare the total wattage value to the continuous wattage rating of the generator.

Table 4-1. Typical equipment wattages.

Equipment	Typical wattage
Farm equipment	
Bulk milk cooler	1,500-12,000
Electric fencer	7-10
Feed conveyor	800-5,000
Feed grinding	1,000-7,500
Feed mixing	800-1,500
Gutter cleaner	3,000-5,000
Infrared lamp	250
Vacuum pump	800-5,000
Milking parlor heater	2,000-10,000
Shop tools	300-1,500
Silo unloader	2,000-7,500
Space heater	1,000-5,000
Ventilation fans	300-800
Water heater	1,000-10,000
Water pump	500-2,500
Yard light	100-500
Essential home equipment	
Electric heater	600 and up
Freezer	600-1,000
Furnace blower	400-600
Furnace oil burner	300
Furnace stoker	400
Refrigerator	400-800
Optional home equipment	
Central air conditioner	2,000-5,000
Coffeemaker	1,000-1,500
Dishwasher	300 + 1,500 for heater
Electric clothes dryer	500 + 4,000 for heater
Electric fan	75-300
Electric iron	500-1,500
Electric range	3,000-4,000
Electric skillet	1,150-1,500
Kitchen ventilator	150
Mixer	150
Sewing machine	200-500
Sweeper	400-1,500
Television	200-600
Toaster	1,200-1,500
Washing machine	400
Water heater	1,000-5,000
Water pump	800-2,500
Window air conditioner	1,000-2,500

Table 4-2. Motor wattages.

Starting and full-load running wattages required for single-phase electric motors.

Motor (hp)	Watts required	
	To start	To run
1/6	1,000	215
1/4	1,500	300
1/3	2,000	400
1/2	2,300	575
3/4	3,345	835
1	4,000	1,000
1-1/2	6,000	1,500
2	8,000	2,000
3	12,000	3,000
5	18,000	4,500
7-1/2	28,000	7,000
10	36,000	9,000

EXAMPLE

4-1. Sizing a generator.

Size a generator for a swine farrowing building with the following electrical equipment:

Full-load system		
Equipment	Starting (watts)	Running (watts)
Fan #1, 1/6 hp (pit)	1,000	215
Fan #2, 1/6 hp (pit)	1,000	215
Fan #3, 1/4 hp (variable speed)	1,500	300
Fan #4, 1/3 hp (summer)	2,000	400
Heater fan, 1/4 hp	1,500	300
Office heater	–	1,466
Water heater	–	3,000
Refrigerator, 1/4 hp	1,500	300
Heat lamps	–	2,500
Lights	–	2,510

Determine the required wattage ratings for both a full-load and a part-load system. Assume that Fan #1 is the only fan that operates during cold weather and that an extra 10% capacity will be needed in the future.

Continued

SOLUTION
The heater fan and office heater would not run at the same time as Fans #2, #3, and #4, so include only the group with the largest total wattage:

Total starting wattage =
1,000 + 4,500 + 3,000 + 1,500 + 2,500 + 2,510 =
15,010 watts

Add 10% for future expansion:
15,010 x 1.10 = 16,511 watts
(minimum generator size for starting wattage)

The running wattage without the heater fan and office heater is:

Total running wattage =
215 + 215 + 300 + 400 + 3,000 + 300 + 2,500 + 2,510
= 9,440 watts

The part-load system assumes that you can get by without fans #3 and #4, the office heater, the water heater, and all lights except the alley lights.

The critical loads and their wattages are:

Equipment	Starting (watts)	Running (watts)
Fan #1, 1/6 hp	1,000	215
Fan #2, 1/6 hp	1,000	215
Heater fan, 1/4 hp	1,500	300
Refrigerator, 1/4 hp	1,500	300
Heat lamps	–	2,500
Alley lights	–	600

The restart sequence and corresponding wattages are:
1. **Heater fan:** 1,500 watts
2. **Fan #1:** 300 + 1,000 = 1,300 watts
3. **Fan #2 (if needed):**
   300 + 215 + 1,000 = 1,515 watts
4. **Refrigerator:**
   300 + 215 + 215 + 1,500 = 2,230 watts
5. **Heat lamps:**
   300 + 215 + 215 + 300 + 2,500 = 3,530 watts
6. **Alley lights:**
   300 + 215 + 215 + 300 + 2,500 + 600 = 4,130 watts

The maximum wattage from these six steps is 4,130 watts. Add 10% for future expansion:

4,130 x 1.10 = 4,543 watts
(minimum generator size for running watts)

Note: This total wattage when the alley lights are turned on (4,130 watts) is equal to the total running wattage.

Installation

Install wiring and equipment to meet *NEC* requirements, local regulations, and the requirements of the power supplier.

Single-phase standby generators are connected to the electrical line by a double-pole, double-throw transfer switch, Figure 4-2, *NEC 702.4*. This prevents accidentally feeding power back into the utility lines to the neighbors or to workers servicing the lines and protects the generator from damage when power is restored.

If the switch is outside, it must be located in a watertight box and properly grounded. Size the switch according to the rating of the farm service entrance. Common sizes are 100, 200, or 400 amps. If the standby generator starts automatically, the automatic switch in the control panel may act as the transfer switch.

A common location for this switch is on the central meter pole. Install the switch between the watt-hour meter and the service disconnect (main fuse box) for the farm. Note that the white (neutral) conductor is usually not switched, but some power suppliers require it to be switched also. When the handle is up, the utility black and red conductors are connected to the load black and red conductors, respectively. In the down position, the load conductors are disconnected from the utility conductors and connected to the black and red conductors from the generator.

A portable, tractor-driven generator does not require special facilities, but store it inside when not in use. If it is to be permanently mounted at the meter pole, anchor it on a 6-inch thick concrete pad, and shelter it from the weather. Provide adequate ventilation in the shelter so the generator does not overheat, and provide easy access to the equipment for servicing. Mark a floor guide for the tractor to help you quickly and accurately position and align the tractor.

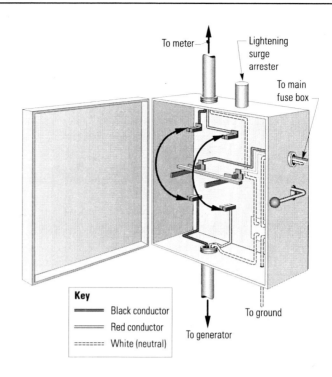

Figure 4-2. Double-pole, double-throw transfer switch.

Store the PTO shaft with the generator. Maintain safety shields on the shaft. If the generator is mounted on a trailer, tightly secure the mounting base to the trailer so it cannot spin around the PTO when under full load. Cover all generator openings with 1/4-inch galvanized wire mesh to prevent damage by rodents. Keep the mesh clean of dust and trash accumulations because they restrict airflow around the generator.

Protect the tractor engine cooling system from blowing snow, dust, or other material that could block the air intake or exhaust. Engines can be ruined by snow blockage of the air cleaner, so an inside location is preferred for blizzard conditions. Store the tractor where it is accessible even in severe weather, when standby power is most likely needed. Do not store the tractor or generator in a building with an electrically powered door. Also, do not depend on an electrically powered pump for tractor fuel.

Locate stationary, engine-driven generators in a safe, dry place within 25 feet and within sight of the transfer switch. In livestock facilities, the engine-driven generator is often placed in a building attached to the livestock structure. This keeps all of the standby equipment indoors, where it is easy to maintain. Mount the generator on a trailer, or bolt it to a permanent, reinforced concrete base at least 6 inches thick. Anchor the mounting bolts in concrete. Protect the generator with a building and provide adequate ventilation. The engine of a 5-kW generator can produce as much heat as a house furnace. Ventilate the engine exhaust fumes to the outside, and provide adequate air intake to the engine.

Other considerations for installing engine-driven generators include clearance for servicing, vibration damping pads, and fuel storage. Manufacturers usually provide detailed installation instructions for their generators; be sure to follow them carefully.

Use a tractor or standby engine with a power output rating of about 2 hp for each kilowatt output of the generator. Generators must be operated at a constant speed, so equip the engine with a tachometer and a governor.

Install power failure alarms for systems with manual start generators and where power outages cannot be tolerated for more than a few minutes. Sometimes an outage can occur in one building and not on the entire farmstead. To be safe, install an alarm in each building where electrical service is critical.

For additional information about generators, contact the Electrical Generating Systems Association (EGSA) at 1650 S. Dixie Hwy, STE 400, Boca Raton, FL 33432; 561-750-5575, or visit their website at www.egsa.org. Your local power supplier can advise you on the generator's proper installation.

Operation

If the generator is automatic, it should start when line power fails. It should stop when or shortly after regular power is restored.

With manual start generators:
1. Turn off or disconnect all electrical equipment
2. Position the tractor for PTO connection, or connect a trailer-mounted engine and generator.
3. Start the engine and bring it up to proper speed (1,800 or 3,600 rpm). The voltmeter shows when it is ready to carry the load. Do not increase generator speed to raise voltage; use the frequency meter. The frequency meter should read 60 Hz, plus or minus 3 Hz at most. Increasing generator speed above 60 Hz does not increase power output and may cause problems for electric motors and other frequency-sensitive equipment.
4. Place the transfer switch in the generator position.
5. Start the motors and equipment in the sequence determined when sizing the generator. If the generator's over-current device shuts off, turn off all the electric equipment and restart.
6. Check the voltmeter frequently. If it falls below 200 volts for 240-volt service or 100 volts for 120-volt service, reduce the load by turning off some equipment.
7. When line power is restored, turn off all the electrical equipment slowly, one load at a time; place the transfer switch into the line power position. Stop the generator, and turn the electrical equipment back on.

Maintenance

Operate the generator under a load of at least 50% of the generator's capacity every month to ensure that it will operate when needed. Check for fuel leaks. Keep the fuel tank full to reduce condensation and to ensure that enough fuel is available when needed. Make the fuel supply large enough to run the generator for several days, because weather conditions may delay fuel deliveries. Follow proper engine maintenance according to the recommendations of the engine manufacturer.

Additional information on maintenance, operation, and safety of standby power systems can be found in ASABE Engineering Practice EP364, "Installation and Maintenance of Farm Standby Electric Power".

CHAPTER 5

Alarm Systems

AN ALARM SYSTEM can warn of electrical failure, equipment failure, unauthorized entry, fire, and/or toxic gases. An alarm system has two major parts:
1. A detector.
2. A warning device.

Some systems have microcomputers and telephone callers with pre-recorded messages; others have just the detector and a switch connected to a light. The simplest devices can be homemade. More sophisticated alarm systems should be purchased. The information in this chapter will help identify some specific tools to use when specifying or reviewing proposals for alarm systems. ASABE Standard S417 *Specifications for Alarm Systems Utilized in Agricultural Structures* has detailed information on understanding the terms used with alarm systems and equipment.

Homemade Alarms

Most homemade alarm systems issue a warning with a light or horn. Figures 5-1 to 5-3 show systems that warn if power to a fan (or fans) fails. Although a fan motor is indicated, any critical motor could be monitored by these systems.

Battery-Operated, Relay-Controlled Alarm

Figure 5-1 shows a magnetic relay in the fan circuit that is energized when there is power to the fan. The contacts of this normally closed relay are a switch for a battery-operated alarm. If power fails, the contacts under spring tension close to sound the alarm. A type of magnetic relay rated at fan voltage for continuous operation with single-pole, single-throw normally closed contacts may be used. Include a test switch (normally closed) unless the fan is connected to a disconnect switch or circuit breaker.

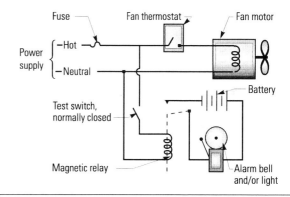

Figure 5-1. Battery-operated, relay-controlled alarm.

Solenoid Valve-Controlled, Compressed Gas Horn

In the alarm system shown in Figure 5-2, an air horn powered by a can of compressed gas, commonly used as a fog horn or distress signal for small boats, is used with a normally open solenoid valve. The solenoid valve stays closed as long as there is power to the fan. With an interruption of power, a spring in the valve opens it, allowing the compressed gas to operate the horn. Again, include a test switch.

Alarm for Multiple Fans

Wiring an alarm system for multiple fans, as shown in Figure 5-5, requires a relay for each fan, but only one alarm.

Battery-Operated Alarm with Thermostat

The alarm systems in Figures 5-1 to 5-3 respond to power failure to the fan. However, a ventilating system may fail for other reasons. Figure 5-4 shows an alarm system that

53

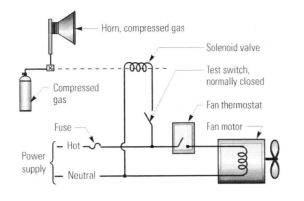

Figure 5-2. Solenoid valve-controlled, compressed gas horn.

responds to heat buildup that normally occurs during a ventilating system failure. This alarm does not sound until conditions are critical.

A cooling thermostat (contacts close on temperature rise) is the sensing element, sounding the alarm at a preset temperature, indicating the air temperature is above normal. In some cases the thermostat should be reset to adjust for seasonal changes—between 85 to 90 °F under summer conditions and lowered to 70 to 75 °F in winter. This system may also indicate when a water sprinkler should be turned on. Test the alarm circuit by temporarily changing the thermostat setting.

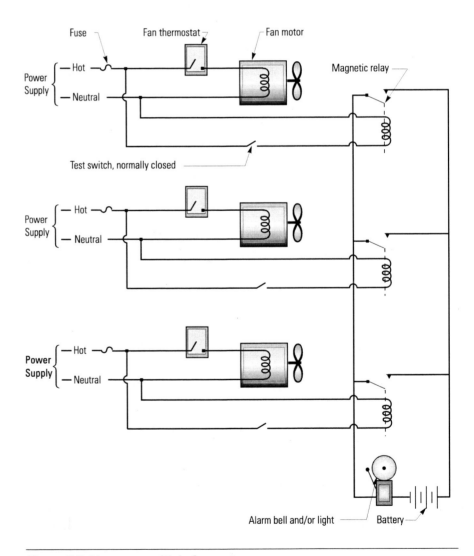

Figure 5-3. Alarm on multiple-fan system.

Combination Alarm System

The alarm shown in Figure 5-5, which combines the circuits shown in Figures 5-1 and 5-4 gives added protection. Failure of ventilation due to broken or slipping belts, plugged air ducts, etc., is detected through increased air temperature, while a power failure to the fans is detected immediately.

Some alarm systems have several alarm circuits like those shown in Figures 5-3 and 5-5 that are wired via an alarm communication circuit to a central telephone autodialer alarm system. Consider placing a surge protection system for the autodialer. The surge protection needs to be in conjunction with a lightning arrestor for the SEP.

Battery Maintenance

Some systems require a battery. Replace dry-cell batteries periodically, and continuously charge wet-cell batteries with a trickle-charger. Check water level in wet-cell batteries yearly. The system shown in Figure 5-2 requires no battery.

Fire Detectors

Most fire detectors sense the smoke buildup in a closed space. There are two types of smoke detectors:
1. Ion.
2. Photo electric.

The ion detector is better for cleaner burning fires, while the photoelectric unit is better for smoky fires. Do not use either type of smoke detector in livestock buildings, because dust in the air causes false alarms.

Heat detectors work more dependably in livestock buildings than do smoke detectors, but proper location is extremely important.

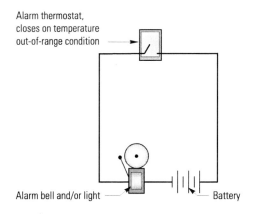

Figure 5-4. Battery-operated alarm with thermostat.

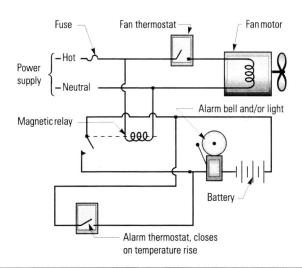

Figure 5-5. Temperature- and power-sensitive alarm.

CHAPTER 5. ALARM SYSTEMS

CHAPTER 6

Stray Voltage

STRAY VOLTAGE is a small voltage difference that exists between two surfaces (stanchion, waterer, floor, etc.) that an animal can touch. When an animal touches both surfaces simultaneously, 60-Hz AC current flows through its body, as illustrated in Figure 6-1. At voltage levels that are just perceptible to the animal, behaviors indicative of perception such as flinches may result with little change in normal routines. At higher levels, avoidance behaviors may result. The first mild behavioral responses for dairy cows in farm-like conditions occur at 2.5 mA of current for the 5% most sensitive cows, 4.8 mA for the 50th percentile, and 8.5 mA for 95% of cows for 60 Hz current measured as root-mean-square (rms) values.

It is clear from several studies as well as physical principles that real-world contact resistances have enormous variability. The lowest contact resistances would be expected if a clean, wet body part (for example, a cow's muzzle) comes into contact with a clean, wet, metallic object with

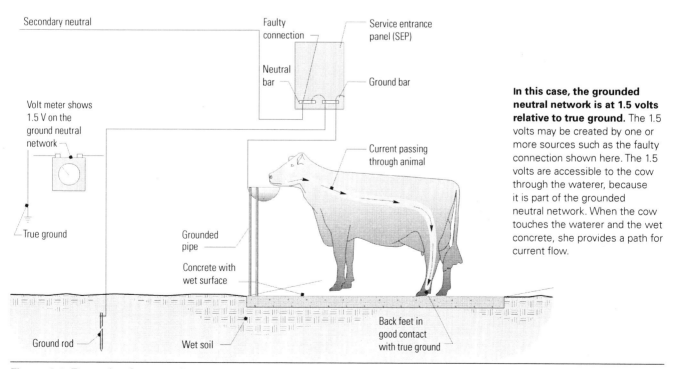

In this case, the grounded neutral network is at 1.5 volts relative to true ground. The 1.5 volts may be created by one or more sources such as the faulty connection shown here. The 1.5 volts are accessible to the cow through the waterer, because it is part of the grounded neutral network. When the cow touches the waterer and the wet concrete, she provides a path for current flow.

Figure 6-1. Example of stray voltage.

a substantial mutual contact area and substantial contact pressure. The accepted practice by researchers and regulators has been to assume worst-case (lowest practical values) for contact resistances. Studies done to measure more typical body + contact resistances that would occur on farms have shown that 500 ohms is a reasonable value to use in a measurement circuit to estimate the current that would flow through a cow's body. Although the resistance of the cow's body is typically less than 500 ohms for the muzzle to hoof pathway (other pathways have a higher resistance), 500 ohms has been shown to be a "worst case" or minimum resistance value for the combination of a dairy cow's body—real-world contact resistance in the farm environment. Contact resistance in dry environments can be several orders of magnitude above these worst-case values.

Regulatory standards (WI, MI, ID) use a value of 2 mA (1 volt in low resistance environments (500 ohms cow + contact) or 2 volts in 'typical' resistance environments (1,000 ohms cow + contact). This presumably considers the low percentage of cows that might perceive currents between 1 and 2 mA as not a problem.

Causes

Stray voltages usually are caused by electrical currents in the grounded neutral network. As shown in Figure 6-2, a common grounded neutral network consists of the following:

- The primary neutral from the power supplier. (Some systems may not have the primary neutral grounded and connected to the secondary neutral.)
- The secondary neutral to the building SEP.
- The branch circuit neutrals.
- The equipment grounds.
- The grounding electrode conductors and ground rods.
- The grounded metal equipment (stanchions, waterers, etc.) in or near the building.

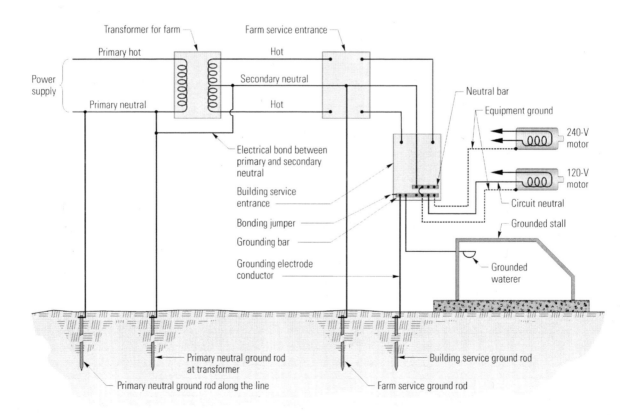

Figure 6-2. Neutral and equipment grounding system.
Single-phase power. Most hot wires have been left off to improve readability.

On-Farm Sources of Stray Voltages

The most common on-farm sources of stray voltages include the following:
- Ground-faults.
- Equipment ground used as a circuit neutral or a circuit neutral used as an equipment ground.
- Unbalanced 120-V loads in the SEP and an undersized secondary neutral or a bad connection in the secondary neutral that causes voltage drop.
- Electric fence wires incorrectly grounded to the electrical grounding system, or short-circuiting to or inducing a voltage in pipes and equipment.

Ground-faults. Electrical current from the farm transformer must return to the transformer. If current can escape from the hot wire, it returns to the transformer through all conductive paths available. The lower the resistance of the path, the more current it conducts. If the hot wire is ground-faulted to a metal object that is connected to an equipment ground, most of the current returns to the transformer via that conductor. The current should reach high enough levels to trip a circuit breaker or blow a fuse. However, without an equipment ground, the current takes a higher resistance path through earth, as illustrated in Figure 6-3. The higher the resistance of this path, the higher the voltage on the equipment with the ground-fault. There probably will not be enough current to open the short-circuit device. Consequently, there is a relatively high voltage between the equipment with the ground-fault and earth. An animal or person that contacts the equipment and earth can receive a shock. Note that the current must return to the transformer via the grounded neutral system. This can lead to stray voltage problems on the system.

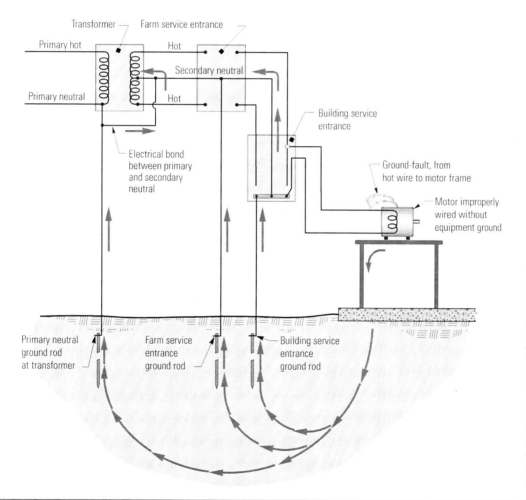

If a path develops for current to flow from the hot wire to earth, the current flows through the earth to the grounding electrodes and back to the transformer. With a properly installed equipment ground, most of the ground-fault current follows the lower resistance path of the equipment ground and secondary neutral back to the transformer, and the circuit breaker will probably be opened.

Figure 6-3. Ground-fault.

One common type of ground-fault results when conductor insulation wears down or a connection loosens so a conductor directly contacts a metal object. If not properly grounded, the metal object has enough voltage to be hazardous. Figure 6-4 shows two installations, one properly wired with an equipment ground, the other improperly wired without an equipment ground. If properly grounded, there is probably enough current to open the short-circuit device, which alerts the operator to the problem. Damp accumulated dirt can be a high resistance path and can create a voltage on metal equipment. If not properly grounded, the metal equipment can have enough voltage to disturb an animal. If properly grounded, the voltage on the metal equipment is near that of earth, so the animal should not feel it, as illustrated in the properly wired installation shown in Figure 6-4. The best defense against stray voltages created by ground-faults is a properly installed grounded network.

a. Properly wired with an equipment ground.

Most of the ground-fault current passes through the equipment ground rather than the cow.

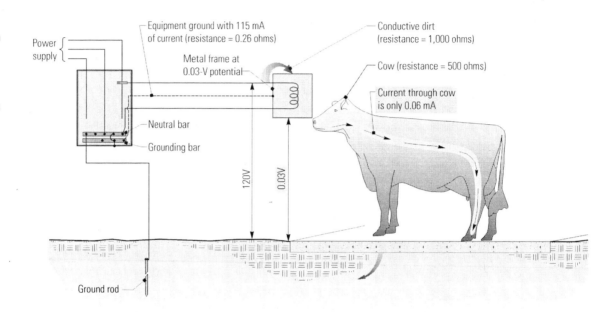

b. Improperly wired without an equipment ground.

The metal frame has a higher voltage, and a higher current passes through the cow without an equipment ground.

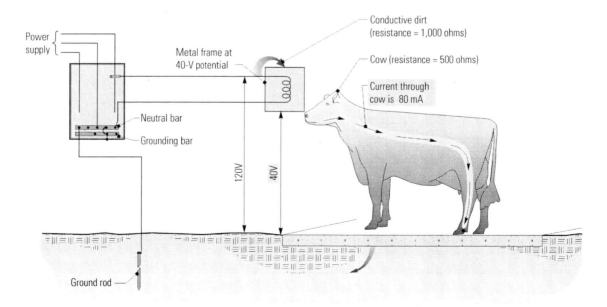

Figure 6-4. Equipment grounding reduces stray voltages.

Another form of ground-fault occurs when an underground conductor is damaged and faults to the earth. This is particularly a problem when underground cable is buried less than 2 feet deep or is not properly protected. Current flows through the highly resistant earth and creates a voltage difference between the earth near the wire and the earth farther away. If an animal contacts the earth near the cable with one leg and the earth farther away with another leg, the animal completes a low resistance path and current flows through it, as shown in Figure 6-5. Faults to the earth can also produce a voltage at a problem level between the grounded neutral system of the farm and the earth. Proper underground cable, careful installation, and burying the cable at least 2 feet deep reduces the risk of this problem, *NEC 300-5*.

Improper interconnection of a circuit neutral and an equipment ground can create stray voltages. *NEC 250.24(A)(5))* requires that these conductors be kept separated throughout all feeder and branch circuits. Circuit neutrals and equipment grounds can be directly connected **only** at the building's SEP. The circuit neutral normally carries current, and if it is connected to the equipment ground away from the SEP, a voltage could develop on the equipment ground and any metal object to which the equipment ground is connected. Figure 6-6 illustrates an improper connection of a circuit neutral and equipment ground.

Unbalanced 120-volt loads in the SEP and poor connections or inadequately sized conductors on the secondary neutral can create a voltage between grounded metal equipment and the earth or floor. The hot wires from the farm transformer typically have a potential difference of 240 volts. The center conductor (secondary neutral) is grounded at the transformer and again at the building's SEP. There is a voltage potential of 120 volts from each of the hot conductors to the secondary neutral. If the current passing through the two hot conductors is not the same, due to unbalanced 120-volt loads at the SEP, then the secondary neutral carries the difference. Figure 6-7 depicts this condition. The better the 120-volt loads are balanced, the less current the secondary neutral carries and the lower the

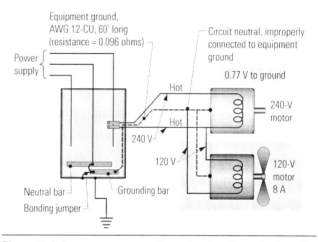

Figure 6-6. Improper connection of a circuit neutral and equipment ground.

A 240-volt circuit was originally wired to a motor. Later a tap was made onto the 240-volt circuit to supply a 120-volt load, and the original equipment ground conductor was used as the circuit neutral of the 120-volt circuit. The circuit neutral normally carries some current, so the equipment ground and the motor frame were charged with enough voltage to shock an animal. If the equipment ground had a poor connection, this voltage would be even higher.

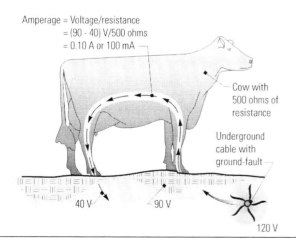

Figure 6-5. Ground-fault from an underground cable.
The voltage is high near the ground-fault and decreases with distance. If an animal completes a low resistance path across this voltage difference, much of the current flows through the animal.

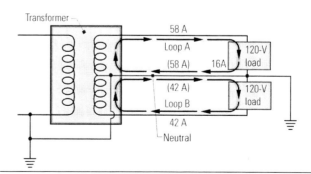

Figure 6-7. Unbalanced 120-volt loads create current on the secondary neutral.
With 58 A on one loop and 42 A on the other, there is 16 A neutral current.

CHAPTER 6. STRAY VOLTAGE

potential for stray voltages. If the secondary neutral has a high resistance due to a poor connection or an inadequately sized conductor, then a voltage difference can be created between the earth or floor and the metal equipment bonded to the neutral at the SEP.

The best way to prevent these problems is to balance the 120-volt loads at the SEP and properly wire the secondary neutral. For livestock buildings, do not use a reduced size neutral conductor. Use the same size as the hot conductors. The first step in balancing the SEP is to use 240-volt motors in place of 120-volt motors. Motors operated at 240 volts do not draw current on the secondary neutral. Also, try to connect the same size and number of 120-volt circuits to each hot conductor in the SEP. For example, if two high load and two low load 120-volt circuits are connected to a SEP, install one high and one low load circuit on the left hot conductor, and do the same on the right hot conductor. Figure 6-8 shows balanced 120-volt loads at a service entrance panel.

Electrical fence wires can cause stray voltages by faulting to ground through a piece of equipment, by improper grounding of the fence, or by inducing a voltage in a pipe running near and parallel to the wire. The high voltage pulse of electrical fence controllers exceeds the 600-volt rating of building electrical wire, so look for insulation breakdown as a possible source of a ground-fault. An electric fence controller must have its own separate ground rod. This ground rod **must not** be connected to the grounding electrodes of the building. When possible, locate the electric fence controller outside of the building. Induced voltage in pipes can be eliminated by grounding the pipe.

Off-Farm Sources of Stray Voltages

For the most common type of distribution system, the grounded wye system, the secondary neutral on the farm is connected to the utility's primary neutral, as shown in Figure 6-2. This bond is made to protect the customer in case there is damage to the power lines or transformer. However, this bond can introduce off-farm stray voltages to the farm. Off-farm sources of stray voltage can be created by

- Voltage on the primary neutral.
- A ground-fault at a neighbor's farm.
- Problems with the off-farm electrical distribution system.

In some cases, a second type of primary distribution system, a three-phase delta system, may be available. In these cases no load to a primary neutral exists, thereby reducing off-farm sources.

Many farms are served with a single-phase distribution line that has only two wires, one of which is grounded. If a large load is served, an unacceptable voltage difference can be created between the primary neutral and earth. The voltage on the primary neutral can lead to some voltage difference between earth and the equipment bonded to the secondary neutral via the primary and secondary neutral bond. A typical single-phase branch originates from a three-phase, four-wire wye primary. The utility can balance the loads on the four-wire primary to reduce the current on the neutral. However, since the utility cannot control the loads the customers use, the balancing can only reduce, not eliminate, the primary neutral current.

As mentioned earlier, when a ground-fault occurs, the current returns to the transformer. Some of the ground-fault current may find a path back to the source transformer via the grounding system at a neighboring farm. Figure 6-9 illustrates how a ground fault at one farm can create a problem at a neighboring farm. Some reports state that this stray voltage source could be one mile or more away and still affect livestock. Correct the ground-fault at the neighboring farm to solve the problem.

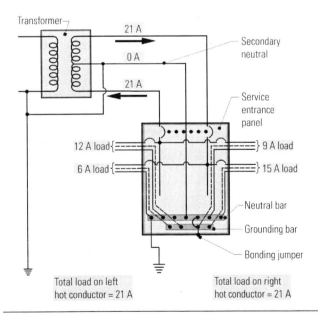

Figure 6-8. Balanced 120-volt loads at the service entrance panel.
Because there are 21 amps on each hot conductor, the secondary neutral carries no current. A perfect balance is rarely achievable, but try to balance as much as possible. Any 240-volt circuit draws equal current from both hot conductors, so it never contributes to unbalance.

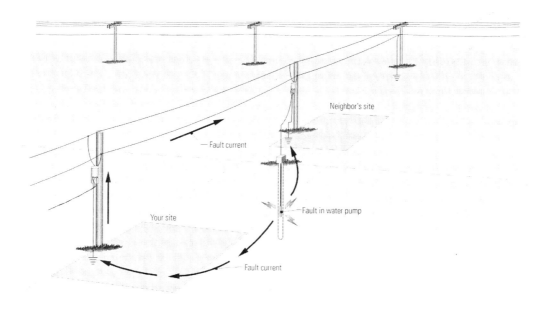

Figure 6-9. Ground-fault at a neighboring farm.
Some of the ground-fault current may return to the source transformer via the grounding electrode of an adjacent farm.

There have been cases where stray voltage has been traced to the breakdown of electrical equipment in the off-farm distribution system. Lightning may damage a transformer, regulator, capacitor, or other equipment. There may also be a poor connection in the primary neutral. The utility may have to completely check its distribution system serving the farm to detect the source.

Farms near a substation may act as a return path for some of the primary neutral current that is returning to the substation. Some of the primary neutral current returns through the earth. The farm grounding system may have a resistance low enough to cause some of the primary neutral current to return to the substation transformer via the farm, as shown in Figure 6-10.

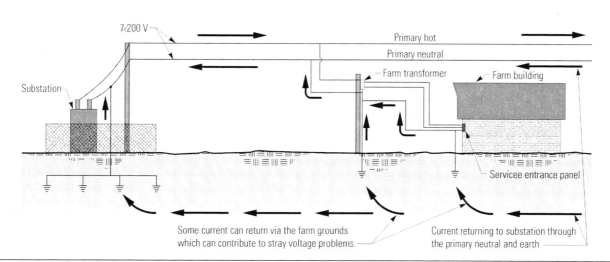

Figure 6-10. Primary neutral current returning to substation via a nearby farm.

CHAPTER 6. STRAY VOLTAGE

Solutions

The solution can be simple if the problem source or sources are clearly diagnosed and the alternatives evaluated. Good diagnosis requires a thorough understanding of electricity and farmstead wiring fundamentals. Often there is more than one cause, which further complicates the diagnosis. If you feel that there may be a problem, contact your power supplier and electrician to diagnose it. It is to everyone's benefit if the farmer, electrician, power supplier, equipment supplier, and others involved work together to attack this problem.

There are three basic solutions to stray voltage problems:
- Eliminate or reduce the voltage causing the problem.
- Isolate the voltage from any equipment in the vicinity of the animals.
- Install an equipotential plane to keep all possible animal contact points at the same voltage.

Eliminate or Reduce the Stray Voltage

If diagnosis indicates that current on the primary neutral system is a major contributor, due to either on-farm or off-farm loads, ask the power supplier to check the distribution system.

If the diagnosis indicates that a voltage drop on the secondary neutral system is a major contributor
- Make sure all connections of the circuit neutrals and equipment grounds are uncorroded and tight.
- Make sure all the circuit neutrals and equipment grounds are large enough in diameter and not excessively long.
- Balance the 120-volt loads from the SEP as much as possible.
- Change 120-volt motors to 240-volts.

Correct the wiring problem if the diagnosis indicates major contributions from
- Currents on the equipment ground.
- Improper interconnection of circuit neutrals and equipment grounds. Keep the two wires separate throughout all branch circuits. These conductors can be connected together only at the neutral bar of the SEP.

To avoid these problems, install the secondary neutral and equipment grounding system strictly by the requirements in the *NEC*, as outlined in this handbook. See Chapter 3, Service Entrance

If the farmstead system contains long secondary neutrals, use a four-wire service to the building as allowed by the *NEC547.9(C)*. Figure 3-5 shows a schematic of the four-wire system to a sub panel in another building, which can reduce the contribution of the secondary neutral voltage between the main service and outbuilding panels to the grounding system at the outbuilding. Only the neutral-to-earth voltage on the grounding system at the main service will be carried to the outbuilding ground system by the fourth wire (the grounding conductor). Neutral-to-earth voltages will remain from all other sources. For the four-wire system, all neutral and grounding conductors in the outbuilding service and all feeders from that service must be completely separated. The originating end of the four-wire system must meet all *NEC* requirements as a service, that is, disconnect with short-circuit protection, system grounding, appropriate enclosure, and neutral-to-ground bonding connection.

Isolate Voltage

If the diagnosis shows a major contribution from the primary neutral, it is possible to isolate this voltage from electrically grounded equipment near livestock. One option is for the power supplier to disconnect the bond between the primary and secondary neutrals at the distribution transformer. Although the *National Electric Safety Code* (*NESC*) allows this, some power suppliers prohibit it because of safety considerations. Power suppliers, if they agree to do so, can modify the farm distribution transformer by placing an isolating device between the primary and secondary neutrals. Figure 6-11 illustrates this solution. The isolating device acts like a switch that is open under normal operation to reduce stray voltage problems, but closes if high voltage (only about 10 volts or more) is sensed between the two neutral systems.

Another option is to install an **isolation transformer** (240-volts to 240/120-volts) between the distribution transformer and the service entrance of the livestock building. This method is the most expensive to apply since the transformers are quite expensive, and they must be mounted more than 8 feet above the ground. This results in extra labor and equipment costs. Isolation transformers are also energy users, as the magnetizing current flows continuously and can easily cost tens of dollars per month.

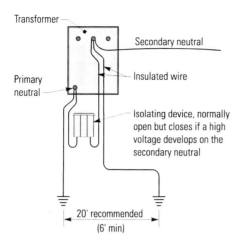

Figure 6-11. Distribution transformer with isolating device between the neutrals.

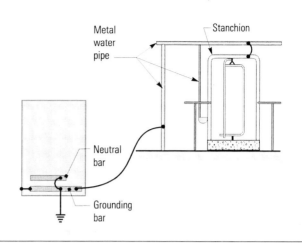

Figure 6-13. Grounding non-electrified metallic equipment.

The isolation transformer can be at either the main farm service, shown in Figure 6-12a (page 66), or at the barn service entrance, shown in Figure 6-12b. If the isolation transformer is at the barn service entrance, it also reduces currents in the secondary neutral due to unbalanced 120-volt loads at the SEP. Care must be taken in proper installation to meet prevailing codes and recommendations, particularly for short-circuit protection, bonding, and grounding.

To completely isolate the voltage, remove any above- or below-ground conductors that may bypass the isolation. Some common interconnections are telephone ground wires, metal water and propane pipes, metal buildings and feeding equipment, and fences between buildings. **Any** conductor reduces the effectiveness of isolation—test to verify the absence of metal interconnections.

Do not try to solve your stray voltage problem by isolating metal equipment (for example, pipes, stanchions, etc.) from the electrical grounding system, as seen in Figure 6-13. **This can create an electrical hazard.** Any electrical fault to ungrounded equipment can create lethal circumstances. Electrically ground all conductive equipment to the service entrance with an equipment ground, particularly if there is electrical equipment in the area.

Install Equipotential Planes

Starting with the 1996 code, *NEC* required an equipotential plane in areas of livestock confinement. This practice was suggested for dairy facilities prior to 1996. Such an installation reduces risk of the problems of stray voltages in the animal confinement areas and should prevent electrocutions in the event of equipment ground faults. Some buildings or areas that do not have accessible electric equipment do not need equipotential planes. Holding areas for milking parlors need special consideration. Details of constructing equipotential planes are shown in Figures 6-14 through 6-17, and can be found in ASABE Engineering Practice EP473 *Equipotential Plane in Livestock Containment Areas* and in ASABE EP342 *Safety for Electrically Heated Livestock Waterers*.

An effective equipotential plane eliminates stray voltage problems regardless of the source. The *NEC* now requires them for new construction in all areas where electrically grounded equipment is near livestock, as would be the case in milking parlors *NEC 547.10*. Gradient ramps are not required, but the ASABE Engineering practice describes how to construct them if needed.

CHAPTER 6. STRAY VOLTAGE

a. Whole farmstead isolation.
Transformer isolating a whole farmstead.

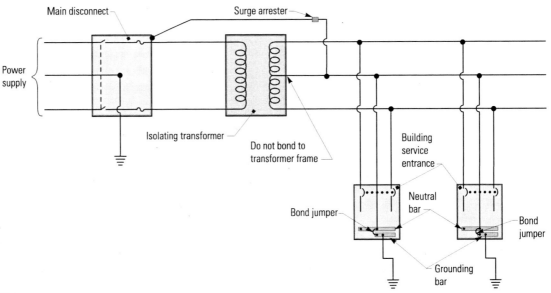

b. Single service isolation.
Transformer isolating a single service entrance.

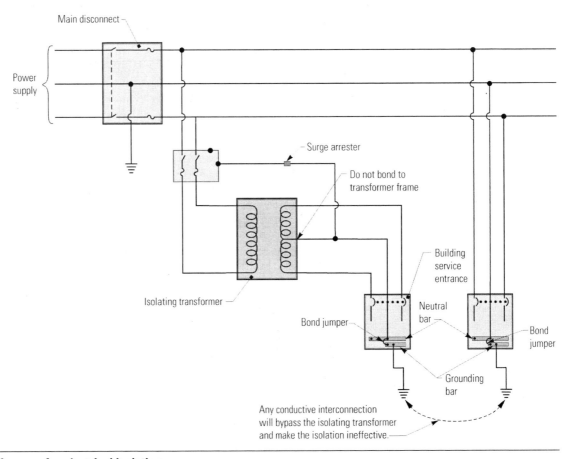

Figure 6-12. Transformers for electrical isolation.
The frame of the isolating transformer is not grounded, because doing so could lead to a dangerous condition. Make the transformer inaccessible by locating it out of reach of people (at least 8 feet above the ground). Alternatives may be acceptable to local authorities. Check with the local authorities before installing an isolating transformer.

a. Waterer supplied by building service equipment.

b. Waterer supplied by its own service equipment.

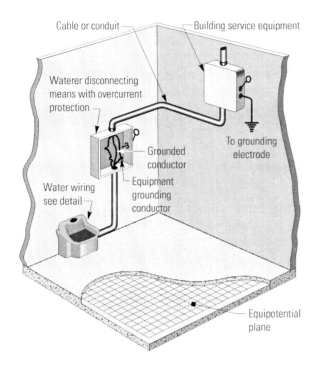
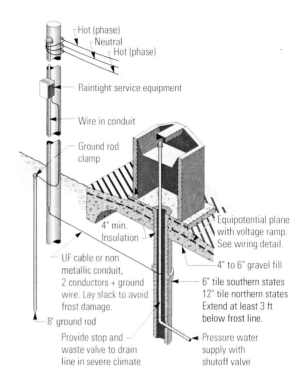

c. Waterer wiring detail.

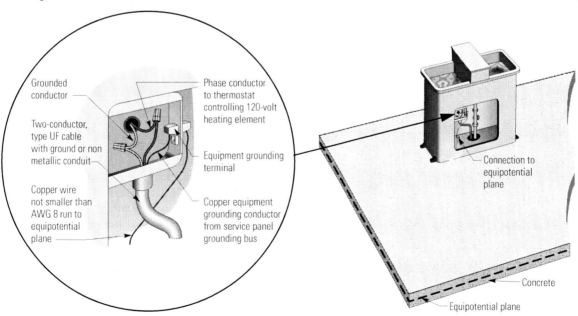

Figure 6-14. Grounding waterers.
Adapted from *Safety for Electrically Heated Livestock Waterers*, EP342.2, The American Society of Agricultural and Biological Engineers, 2950 Niles Road, St. Joseph, MI.

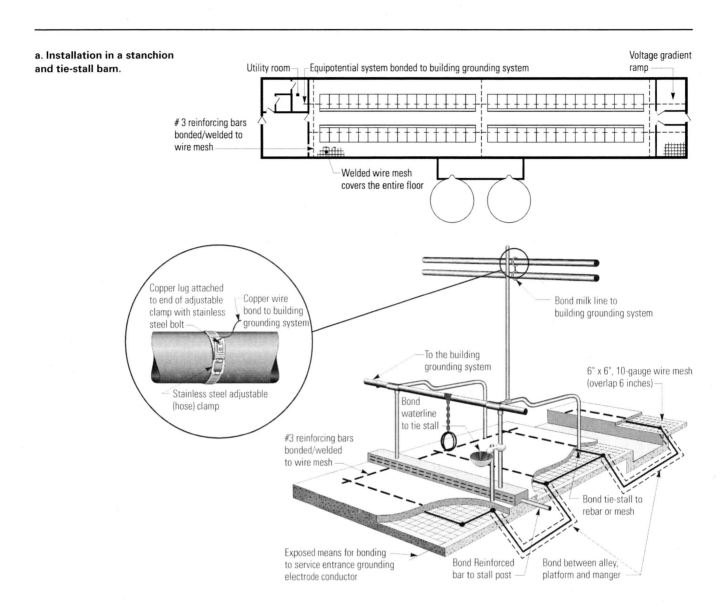
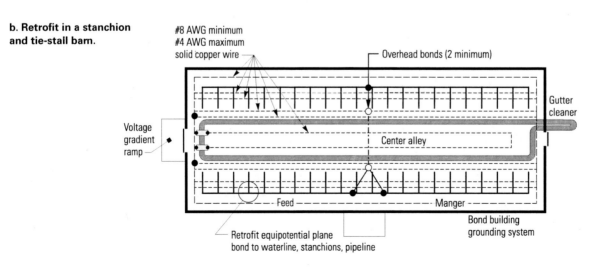

Figure 6-15. Equipotential plane in stanchion and tie-stall barns.
Adapted from *Equipotential Plane in Livestock Containment Areas*, EP473.2, The American Society of Agricultural and Biological Engineers, 2950 Niles Road, St. Joseph, MI.

a. Installation in a milking parlor.

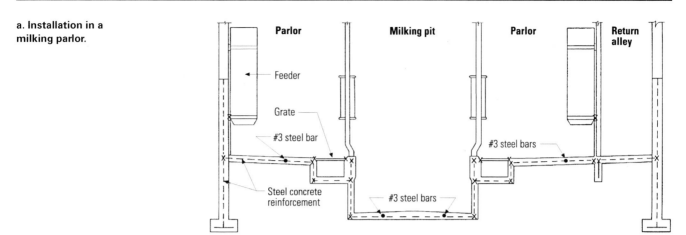

x = Conductive bond. Welding preferred, approved grounding connectors acceptable.

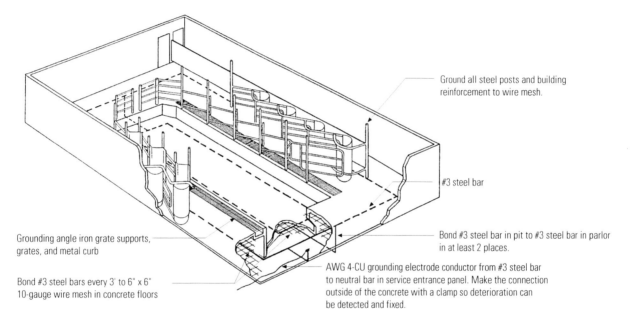

b. Retrofit in a milking parlor.

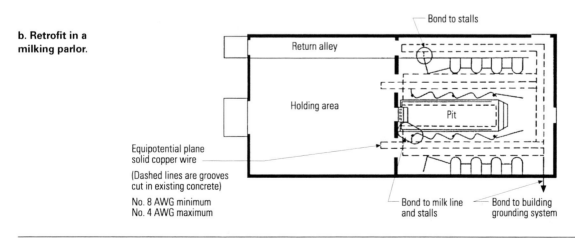

Figure 6-16. Equipotential plane in milking parlor.
Adapted from *Equipotential Plane in Livestock Containment Areas*, EP473.2, The American Society of Agricultural and Biological Engineers, 2950 Niles Road, St. Joseph, MI.

a. Partially slotted floor.

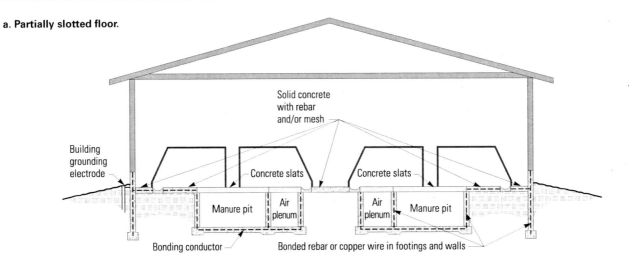

b. Fully slotted floor.

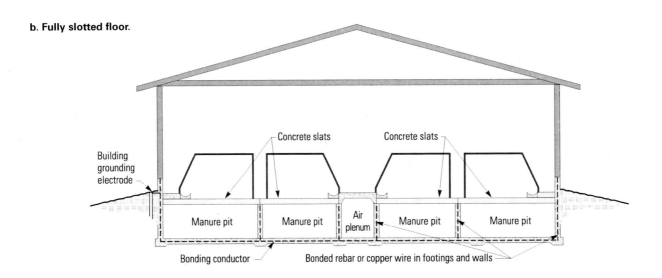

c. Slotted flooring details.
Modified from Pork Industry Handbook fact sheet on Swine Farrowing Units.

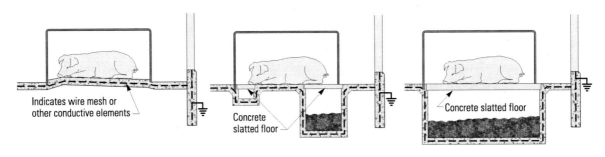

Figure 6-17. Equipotential planes in swine buildings.
Adapted from *Equipotential Plane in Livestock Containment Areas*, EP473.2, The American Society of Agricultural and Biological Engineers, 2950 Niles Road, St. Joseph, MI.

CHAPTER 7

Lightning Protection

FARMERS, FARM OWNERS, and building contractors need a working knowledge of the principles of lightning protection so they can determine their lightning protection requirements, discuss the plans with installation experts, survey the completed installation systematically, and make periodic inspections of the installation. **However, lightning protection systems should be planned and installed only by persons who have the necessary training and equipment. For the best lightning protection, obtain a Master Label System as approved by the Underwriters Laboratories, Inc.**

Lightning is a major cause of farm fires and electrical equipment failures. Electronic equipment, barns, loafing sheds, and other livestock buildings are particularly vulnerable. Damage to farm buildings costs tens of millions of dollars every year. Appliance and equipment losses total additional millions. Lightning causes more than 80% of all accidental livestock losses.

For protection of electronic equipment in buildings without a lightning protection system, see the grounding recommendations in Figure 3-8.

Lightning protection can prevent or greatly reduce this danger to life and property. A properly designed, installed, and maintained lightning protection system affords almost 100% protection to a building. In most states, lightning protection reduces the cost of fire insurance on buildings.

Lightning is electric current with tremendous voltage and amperage. A typical lightning discharge will have 10,000,000 to 100,000,000 volts and 1,000 to 300,000 amps. The temperature of the surrounding air often reaches 50,000 °F, which is four times the temperature of the sun's surface. Lightning voltage is so great that a bolt may leap a mile or more through the air.

Lightning strikes trees, buildings, or other objects because the materials in them are better electrical conductors (easier paths to ground) than air. Lone trees and farm buildings are prime lightning targets because they provide the only ladder in the area for the ground charges to climb nearer to the cloud charges.

Protect all important farm buildings against lightning. Minor buildings, trees and wire fences may also require protection.

Lightning can enter a building in four ways:
- Directly striking a building.
- Striking a metal object, such as a television antenna, cupola, or track extending from the building.
- Striking a nearby tree and leaping over to a building to find a lower resistance path to ground.
- Striking and following a power line or an ungrounded wire fence attached to a building.

The bolt generally follows a metallic path to ground. At points along that path, the main bolt may leap from the wiring to the plumbing, or parts of the current may side-flash to appliances, electronic equipment, water lines, a person, or an animal.

Lightning protection systems are designed to provide a direct, easy path for the bolt to follow to ground and to prevent damage, injury, or death as the bolt travels that path. Protect all points that a bolt is likely to strike and all objects to which the current might side-flash.

A lightning protection system consists of five principal parts:
- Air terminals (lightning rods).
- Main conductors.
- Secondary conductors.

- Lightning arresters.
- Ground connections or rods.

Secondary service arresters and arresters for television antenna lead-in wires are highly recommended for additional protection.

Air Terminals

Air terminals are pointed metal rods or tubes that are installed at high points and other strategic areas of the building, as illustrated in Figure 7-1. Their purpose is to take any lightning bolt that may strike in the immediate area.

Terminals are usually 10 to 24 inches high. The contour and slope of the roof determine the terminal length. Shorter lengths are less conspicuous.

Air terminal diameters range from 3/8 to 5/8 inches. Smaller terminals are usually solid copper, while larger terminals are tubular aluminum.

Space terminals no more than 20 feet on-center along roof ridges, railings, and parapets. Install a terminal within 2 feet of each gable end of a roof.

A flat or low-slope roof requires air terminals 20-feet on-center along the perimeter. A low-slope roof is one with a 3/12 slope for buildings 40 feet wide and narrower or a 6/12 slope for buildings more than 40 feet wide.

All chimneys, dormers, ventilators, flagpoles, towers, water tanks, and other projections need one or more terminals. Silos and towers with flat roofs require two or more.

Main Conductors

The main conductors connect air terminals with each other and with grounds. Their purpose is to conduct the lightning bolt safely from the point struck to the ground. All secondary conductors are connected to them.

Air terminals on each building require at least two down conductors to the ground.

> **Exception:** Conductor drops from a higher to a lower roof level are permitted with one down-lead, provided there are no more than two ridge- or three flat-roof air terminals on the lower level.

Ground each down conductor separately. Buildings with perimeters of more than 200 feet require one additional down conductor with ground for each additional 100 feet of perimeter or fraction thereof.

Separate the down conductors as much as possible; place on diagonal corners of rectangular buildings and on diametrically opposite sides of cylindrical structures.

Use copper or aluminum cable conductors. Figure 7-2 shows three common conductors. Galvanized steel is not recommended. Use copper cable where salt air is common; aluminum corrodes. Placing copper conductors on aluminum roofing or siding causes corrosion. The conducting capacity of a conductor depends on its weight. Minimum weight for copper is 187.5 pounds per 1,000 feet and 95 pounds per 1,000 feet for aluminum.

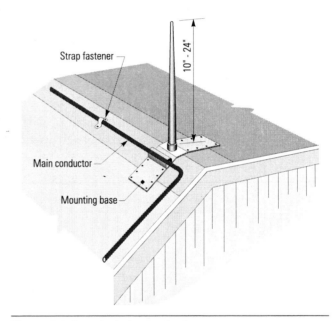

Figure 7-1. A typical air terminal.

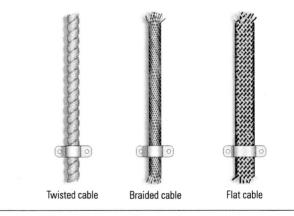

Figure 7-2. Common conductors.

Install conductors as straight as possible between the terminals and the ground, and always maintain a vertical or downward path for the lightning discharge to follow. Avoid bends if possible.

As Figure 7-3 shows, make unavoidable bends at least 90 degrees, and have at least an 8-inch radius.

Join conductors with noncorrosive fittings that provide continuous electrical connections without soldering. Whenever dissimilar metals are joined, use bimetal connectors to prevent corrosion. Fasten conductors with strap fasteners or with special masonry anchors spaced 3 feet apart.

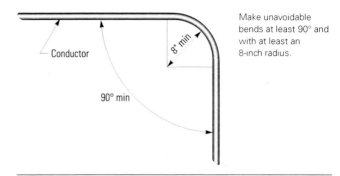

Figure 7-3. Conductor bends.

Secondary Conductors

Secondary conductors connect metal parts of a building together to provide an electrical bond to keep lightning from flashing across the gap between the metal parts. These parts are divided into bodies of conductance and bodies of inductance.

Bodies of conductance are metal objects that are attached to or are part of the building and that may be subject to direct lightning discharges because of their exposure to lightning or their proximity to the lightning conductor. They may include metal hay tracks, high eave troughs, roofing, ventilators, chimney extensions, television antennae, wire fences, clothes wires, and guy wires. Figures 7-4 and 7-5 depict buildings with these features. Unless these metal objects are in direct contact with the lightning conductor or otherwise adequately grounded, connect them to the lightning conductor system. If they are not connected, lightning could flash across the gap.

Bodies of inductance are metal objects near a lightning conductor that may at times build up a charge opposite to that of the grounded lightning conductor. If 6 feet or less from the lightning conductor, they may induce a spark across the intervening gap. Connect them to the lightning conductor system. Metal bodies of inductance may include door tracks, storage tanks, low eave troughs, and downspouts. Interior bodies may include stalls, stanchions, milking lines,

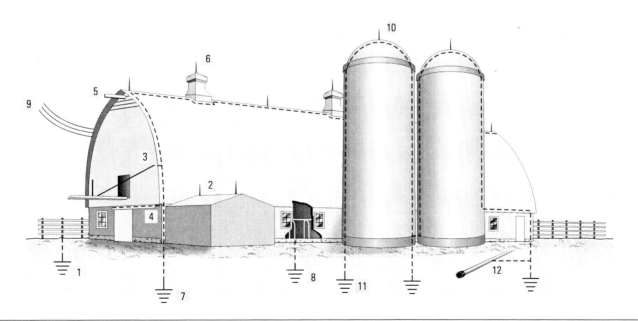

Figure 7-4. Lightning protection for barns.
(**1**) Ground-attached wire fence; (**2**) extend system to an addition; (**3**), (**4**), and (**5**) run secondary conductors to litter, metal door and hay tracks; (**6**) install terminals on cupolas, ventilators, etc.; (**7**) provide at least two grounds for barn; (**8**) tie in metal stanchions; (**9**) install arresters on overhead wires; (**10**) provide at least one terminal on each domed silo and at least two on each flat-roof or unroofed silo; (**11**) ground silo; (**12**) connect to metal water pipes; and protect all buildings—valuable or not—within 50 feet of barn.

Figure 7-5. Lightning protection for houses.
(**1**) Space terminals no more than 20 feet on-center along ridges and within 2 feet of ridge ends; (**2**) install at least two down lead conductors with (**3**) grounds, at least 10 feet deep, for house—additional grounds for clotheslines, etc.; (**4**) tie roof projections such as ornaments into conductor system; (**5**) protect trees within 10 feet of house—connect to house grounding; (**6**) install at least two terminals on chimneys; (**7**) install terminals on dormers; (**8**) install an arrester on antenna—connect to main conductor; (**9**) tie in gutters; and (**10**) make sure there is an arrester on overhead power lines.

water lines, long I-beams, pipes, litter tracks, gutter cleaners, conveyors, long metal ducts, and concrete reinforcement.

Common grounding has been recognized as the most effective means of eliminating side-flashing between metal bodies of inductance. Bond all grounding electrodes to the lightning protection system, including electric and telephone service grounds and piping systems, water service, gas piping, LP gas piping systems, underground metal conduits, etc. Bridge any section of plastic pipe in metal piping systems with a conductor the same size as the grounding electrode conductor.

The *NEC* requires that grounds for lightning systems must also be bonded to the electrical system grounds *NEC-250.106*. In addition, the *NEC* requires that the telephone, TV, and electrical system grounds must all be made at a common point *NEC 250.50* and *800.100(B)(1)*.

If the lightning conductor has been grounded to a metal water pipe in the building, metal bodies of inductance may be connected to the water pipe system, to the nearest lightning conductor, or to another metal object already connected. Use the shortest path to the lightning protection system.

Arresters

Arresters, like those shown in Figure 7-6, prevent dangerous surges of electricity from entering the building's wiring system when lightning strikes the power lines. A power surge caused by lightning is so fast that ordinary circuit breakers and fuses cannot protect wiring and electrical equipment. Arresters divert power from the hot wire and safely guide it to ground when the voltage on the wire exceeds a given level.

Figure 7-6. Lightning arrester.

Arresters are usually installed at the electrical entrance to a building, on telephone service entrances, and on radio and television antennae. Install lightning arresters between the power circuit and ground where the circuit enters any building served by overhead wires. An interior-type of lightning arrester that connects into the SEP on the power supply side is also available. Some SEP's have breaker-like arrestors as part of the assembly. Connect each arrester to the building's lightning protection grounding system.

Arresters are often relatively inexpensive and are especially valuable when there is expensive electrical equipment to protect such as irrigation, air conditioning, and refrigeration equipment. Lightning protection is also required for solid-state devices such as computers, alarm systems, and sophisticated control panels in grain drying systems and livestock buildings. Additional lightning protection can be provided by plug-in surge protectors for this type of equipment.

Ground Connections

Proper grounds are critical to ensure dissipation of a lightning discharge without damage. The extent of grounding depends on the soil, ranging from two, simple 10-foot ground electrodes for a small building located on deep conductive soil, to an elaborate network of cables and rods or plates buried in soil that is dry or rocky and of poor conductivity.

A ground connection is required for each down conductor in the system. That means, provide at least two ground connections on every building and possibly more on larger buildings.

Use stainless steel, copper-clad steel, or solid copper for ground connections because they resist corrosion. Aluminum corrodes when in contact with the soil. If the down conductor is aluminum, connect it to a copper or stainless steel conducting cable with a bimetallic connector. Keep these connections at least 18 inches above ground level. Install wood, metal, or plastic guards at least 6 feet high to protect down conductors from livestock and equipment.

Common grounding methods are shown in Figures 7-7 to 7-9, and Figure 3-8.

Bond all grounding conductors together. This includes electric and telephone service grounds, antenna grounds, all underground metal pipes, and lightning protection systems. Grounding conductors located within 6 feet of each other must be bonded together to avoid side flashes.

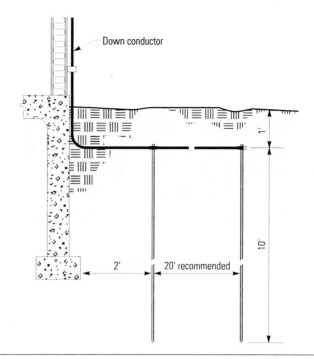

Figure 7-7. Ground rods.
In most clay soils, grounding is accomplished by clamping the conductor cable to a ground rod driven at least 10 feet into the ground. For sandy soils, connect 2 or 3 ground rods in parallel. Space them at least 6 feet apart with a preferred distance of 20 feet and drive them 10 feet deep. Make connections between ground rods and conducting cables at least 2 feet from the foundation wall and 12 inches below ground level. This prevents damage to the foundation from pressures created by lightning dissipation in the ground.

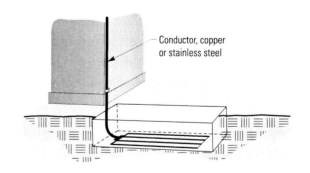

Figure 7-8. Grounded stranded cables.
Grounding in shallow topsoil can be done by stranding the copper or stainless steel conductor and burying it in a trench extending away from the building. Trenches 12 feet long and 1 to 2 feet deep are used in clay soils. In sandy soils, trenches must be at least 24 feet long. When a 24-foot trench is not practical, attach a copper ground plate at least 0.032 inches thick with an area of 2 square feet to the cable at the end of the trench.

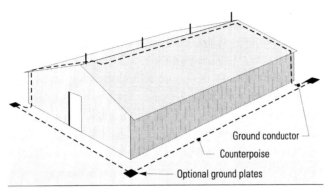

Figure 7-9. Counterpoise ground.
In soils less than 12 inches deep, ground the system by surrounding the entire structure with a conductor not smaller than AWG 2-CU laid in a trench or rock crevice. This is known as a **counterpoise**. Connect the down conductors to the counterpoise. Ground plates are often placed in the corners of the counterpoise.

Metal-Clad and Steel-Framed Buildings

Metal-clad buildings are those with sides and/or roof of sheet metal. Equip these buildings with a lightning protection system as described in the previous section. Do not use metal roofing and siding as main or secondary conductors, but do bond them to the main conductor. Figure 7-10 shows the use of a steel frame as a conductor.

Steel framed buildings may use the steel framework as the main conductor if the framing is electrically continuous and of sufficient size. Sufficient size steel framing will carry the large current without melting or becoming overheated.

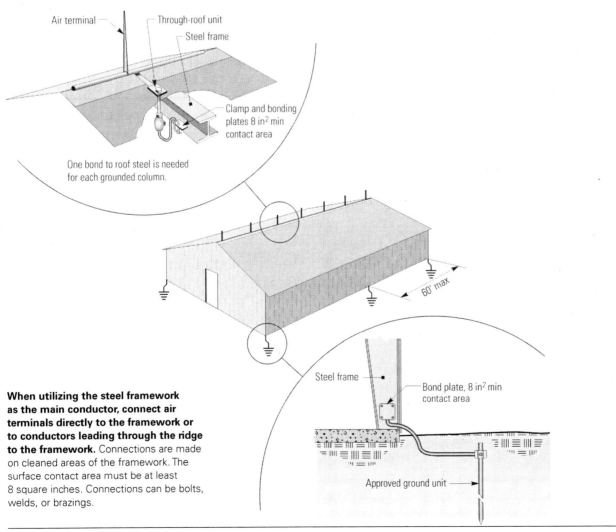

When utilizing the steel framework as the main conductor, connect air terminals directly to the framework or to conductors leading through the ridge to the framework. Connections are made on cleaned areas of the framework. The surface contact area must be at least 8 square inches. Connections can be bolts, welds, or brazings.

Figure 7-10. Steel frame as conductors.

Fences

Improperly grounded wire fences that have been struck by lightning may carry current along the wires for up to 2 miles. Ground fences that are attached to trees or buildings especially those with wooden posts. Wire fences can be grounded in one of two ways:

- Inserting steel fence posts at 150 feet on-center along the fence line, as illustrated in Figure 7-11.
- Driving a galvanized iron pipe beside the wooden fence post and attaching the fence wires with wire ties. Space pipes at a distance of 150 feet on-center. Use a pipe diameter of at least 1/2 inch and drive the pipe into the ground at least 2 feet.

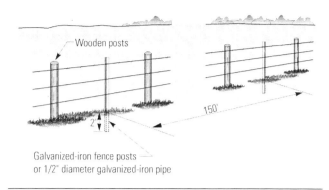

Figure 7-11. Grounding wire fences.

Trees

Trees when struck by lightning pose a threat of falling on buildings and/or livestock. Lightning protection is important for trees within 10 feet of any building, trees that shelter livestock during storms, or trees that are valuable. When trees are located in small groves, only a few of the taller trees need to be lightning protected.

To protect trees, install an air terminal at the top of the trunk (or main branch) and at the end of main branches. Figure 7-12 shows how trees can be grounded. A good connection of the terminals should be installed as far out as possible on the branches. The main conductor of the terminal should run from the trunk to the ground. Connect terminals located on branches to the main conductor.

Locate the ground rod a minimum of 12 feet away from the tree. When the reach of the branches is greater than 12 feet from the tree, locate the ground rod at the edge of the reach of the branches to prevent damage to the root system. Drive a 10-foot ground rod vertically into the ground out of reach of the tree root system. Bury the conductor from the tree to the ground rod in a shallow trench in such a way that the conductor will not be damaged by livestock, and also will be kept away from the root system. To prevent damage, cover any other part of the conductor that will be exposed to livestock.

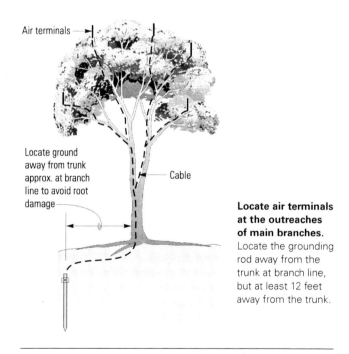

Figure 7-12. Grounding trees.

Locate air terminals at the outreaches of main branches. Locate the grounding rod away from the trunk at branch line, but at least 12 feet away from the trunk.

CHAPTER 8

Example Buildings

THE FOLLOWING EXAMPLES are to be used as guides for designing an electrical system. The lighting levels selected for each example are to meet the needs of the example building. Check with an electrician or contractor to determine light levels for individual needs.

EXAMPLE
8-1. Machine Storage Area

Design the SEP for the machine storage shown in Figure 8-1. Assume the building will be dry and that type **NM** wire (with grounding wire) is acceptable. If conduit is used, the same size of conductors as the **NM** cable would be needed. Use illumination values listed in Table 1-8 as design assumptions and Tables 1-8 and 1-9 for watts per square foot.

SOLUTION

Using Table 1-8, the general machine storage should be illuminated to 5 foot-candles. The number of high-pressure sodium (HPS) light bulbs needed for the building is:

$$\frac{(6 \text{ fixtures})\left(125 \text{ W}/\text{fixture}\right)}{120 \text{ V}} = 6.25 \text{ amps}$$

To provide uniform distribution of light, use six, 100-watt HPS lights as shown in Figure 8-2. Consider wiring some lights on a separate switch so all the lights do not have to be on at the same time.

$$\frac{(40 \text{ft})(104 \text{ft})\left(0.12 \text{ W}/\text{ft}^2\right)}{100 \text{ W}/\text{fixture}} = 4.99 \text{ fixtures}$$

The lights are sustained loads, so increase the current 25% to determine conductor size:

$$1.25 \ (6.25 \text{ A}) = 7.81 \text{ amps}$$

The nearest lights are about 50 feet from the SEP, so use AWG 14-CU conductor in **NM** cable, Table 2-6. The maximum circuit breaker capacity for AWG 14-CU conductor is 15 amps, as listed in Table 3-1.

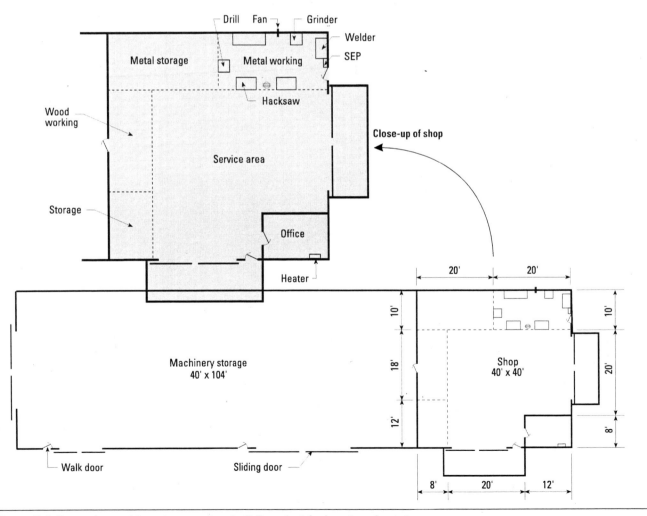

Figure 8-1. Example machine storage shop building electrical system layout.

Machine storage area

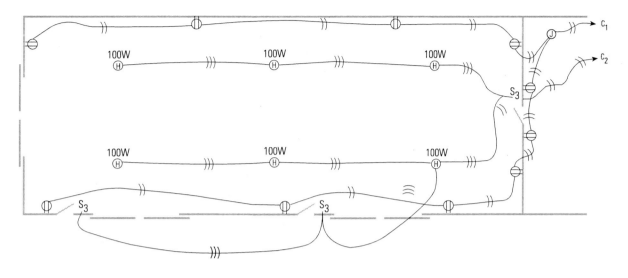

Shop area

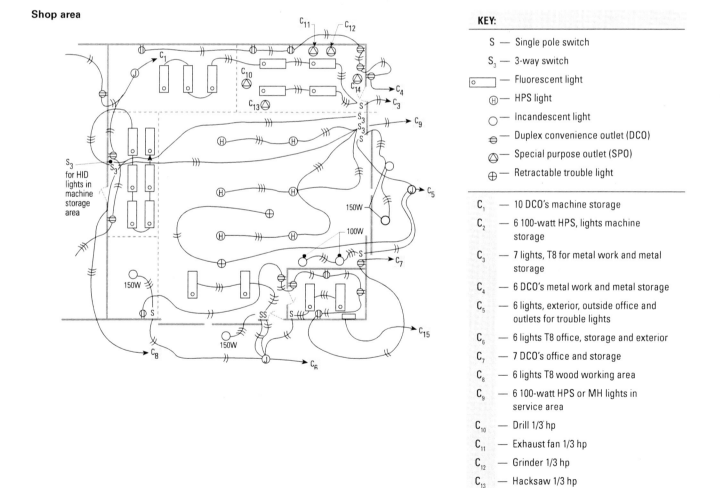

KEY:

S	— Single pole switch
S_3	— 3-way switch
	— Fluorescent light
	— HPS light
	— Incandescent light
	— Duplex convenience outlet (DCO)
	— Special purpose outlet (SPO)
	— Retractable trouble light

C_1	—	10 DCO's machine storage
C_2	—	6 100-watt HPS, lights machine storage
C_3	—	7 lights, T8 for metal work and metal storage
C_4	—	6 DCO's metal work and metal storage
C_5	—	6 lights, exterior, outside office and outlets for trouble lights
C_6	—	6 lights T8 office, storage and exterior
C_7	—	7 DCO's office and storage
C_8	—	6 lights T8 wood working area
C_9	—	6 100-watt HPS or MH lights in service area
C_{10}	—	Drill 1/3 hp
C_{11}	—	Exhaust fan 1/3 hp
C_{12}	—	Grinder 1/3 hp
C_{13}	—	Hacksaw 1/3 hp
C_{14}	—	Welder 240 V, 50 A
C_{15}	—	Office heater 240 V, 5000 Btu

Figure 8-2. Electrical plan for example machine storage building.

Install one DCO about 40 feet on-center along the wall perimeter for a total of eight DCOs. (See Figure 2-3.) All eight of the DCOs can be put on one general-purpose circuit; use AWG 12-CU conductor with a 20-amp circuit breaker.

The SHOP AREA is the next area to design. Start with the storage area. No DCOs are required for this area. An illumination level for storage is not specifically listed in Table 1-8; therefore, choose an activity or task that would most nearly fit the tasks needed around an oil storage area. Feed storage/processing is an activity that nearly matches the activity performed in the oil storage area. The design illumination level is 10 foot-candles. A 150-watt incandescent or halogen incandescent replacement light would work well in this area. The number of light bulbs needed for this area is:

$$\frac{(8\text{ft})(12\text{ft})\left(1.0\,\text{W}/\text{sqft}\right)}{150\,\text{W}/\text{fixture}} = 0.64 \text{ fixtures}$$

Use one 150-watt incandescent or halogen incandescent replacement light with a near-by switch.

The next area to design is THE WOODWORKING AREA. Provide at least two DCOs for power hand tools in this area. Table 1-8 shows the illumination level for a shop bench area to be 100 foot-candles. Fluorescent light fixtures provide good light for woodworking. Using a fixture that uses double, T8 32-watt bulb fixtures (35-watt power usage), the number of fixtures needed is:

$$\frac{(8\text{ft})(18\text{ft})\left(2.75\,\text{W}/\text{sqft}\right)}{(35+35)\,\text{W}/\text{fixture}} = 5.66 \text{ fixtures}$$

Use six, double T8 32-watt fluorescent light fixtures.

The activities performed in the metal storage area will be similar to those in the machine repair area. No DCOs are required. The design illumination level is 30 foot-candles. Using a double, T8 32-watt bulb fixture, the number of fixtures needed is:

$$\frac{(10\text{ft})(20\text{ft})\left(0.83\,\text{W}/\text{sqft}\right)}{(35+35)\,\text{W}/\text{fixture}} = 2.37 \text{ fixtures}$$

Use three, double T8 32-watt fluorescent light fixtures.

The METAL WORKING AREA is a rough bench work area that should be illuminated to 50 foot-candles. Using a double, T8 32-watt bulb fixture, the number of fixtures needed is:

$$\frac{(10\text{ft})(20\text{ft})\left(1.38\,\text{W}/\text{sqft}\right)}{(35+35)\,\text{W}/\text{fixture}} = 3.94 \text{ fixtures}$$

Use four, double T6 32-watt fluorescent light fixtures. Provide DCOs 4 feet on-center along the bench and welding area. Individual branch circuits with SPO are required for the grinder, exhaust fan, drill, welder, and metal bandsaw. All the motors are 1/3 hp (120 V) so their full-load current is 7.2 amps, according to Table 2-3. Motors are sustained loads, so increase the full-load current by 25% to determine conductor size:

$$(1.25)(7.2\text{ A}) = 9\text{ A}$$

All the motors are within 30 feet of the SEP, so use AWG 12-CU conductor in **NM** cable, as listed in Table 2-6. In this example, the motors have no code letter, so the amperage rating of the circuit breaker on each of these individual branch circuits cannot be more than 250% of the motor full-load amperage:

$$(2.5)(7.2\text{ A}) = 18\text{ A}$$

A 15-amp circuit breaker is acceptable and readily available. Each motor is plugged into an SPO and is 1/3 hp. Each motor circuit needs:

- **Short-circuit protection device**—use a circuit breaker in the SEP.
- **Disconnecting means**—All devices except exhaust fans are portable. The welder may be very portable with the cord plug serving as the disconnect. Locate within sight at the DCO.
- **Controller**—Locate within sight. Each device has a built-in switch.
- **Overload protection device**—the motors are manually started, less than 1 hp, within sight and 50 feet of the controller, leave as is so the circuit breaker cannot act as the overload protective device, *NEC 430.32(D)(1)*. The best solution is to use a fused disconnect with 10-amp fuses at each motor. This solution provides the disconnecting means

and overload protection in addition to the plug as a disconnecting means.

An individual branch circuit with an SPO is required for the 50-amp, 240-volt welder. Contact your electrical supplier for the correct type of receptacle. The welder is within 50 feet of the SEP, so use at least AWG 6-CU conductor, as shown in Table 2-6. Protect the conductor with a 50-amp, two-pole circuit breaker, based on values in Table 3-1.

THE SERVICE AREA is a machine repair area that should be illuminated to 30 foot-candles. Using 100-watt metal halide fixtures for best CRI, the number of fixtures needed is:

$$\frac{[(8\text{ft})(20\text{ft})(22\text{ft})(32\text{ft})]\left(0.60\ \text{W}/\text{sqft}\right)}{100\ \text{W}/\text{fixture}} = 5.18\ \text{fixtures}$$

Use six 100-watt metal halide light fixtures. Consider wiring the center row of lights to a separate switch. Add two overhead retractable trouble lights for task lighting. Provide additional lighting over the small storage area and outside aprons. Add two T8 32-watt bulb fluorescent fixtures near the office in the service area. These fixtures will provide immediate light while the metal halide lights come on. It usually takes between 3 to 5 minutes for metal halide lights to fully illuminate; therefore, immediate light is needed.

THE OFFICE should be illuminated to 50 foot-candles. Using a double, T8 32-watt tube fixture, the number of fixtures needed is:

$$\frac{(8\text{ft})(12\text{ft})\left(1.38\ \text{W}/\text{sqft}\right)}{(35+35)\ \text{W}/\text{fixture}} = 1.89\ \text{fixtures}$$

Use two double T8 32-watt fluorescent light fixtures. Provide at least one DCO per wall. An individual branch circuit is required for the 5,000 Btu/hr, 240-volt electric heater. The heater will use:

$$\frac{5,000\ \text{Btu}/\text{hr}}{\left(3.41\ \text{W}/\text{Btu/hr}\right)(240\text{V})} = 6.1\ \text{amps}$$

The heater is within 50 feet of the SEP, so use AWG 12-CU conductor in **NM** cable, as listed in Table 2-6, and protect it with a 20-amp circuit breaker, according to the values in Table 3-1.

The next step is to design the SERVICE ENTRANCE PANEL. All the general branch circuits are arranged so no more than 10 lights or DCOs are connected to one circuit. The general-purpose circuits require AWG 12-CU conductor and 20-amp circuit breakers.

Electrical equipment	Amperage at 240 V
Lights: (37 lights) (1.5 A per light) = 55.5 A at 120 V	27.8 A
Outlets: (23 DCO) (1.5 A per DCO) = 34.5 A at 120 V	17.3 A
Motors (Table 2-3, four 1/3 hp): (1.25) (1.5 motor*)(7.2 A per motor) + (3 motors) (7.2 A per motor) = 30.6 A at 120 A = 34.5 A at 120 V	15.3 A
Welder	50.0 A
Heater: $\dfrac{(1\ \text{heater})\left(5,000\ \text{Btu/hr}/\text{heater}\right)}{\left(3.41\ \text{Btu/hr}/\text{watt}\right)(240\text{V})} = 6.1\ \text{amps}$	6.1 A
Total amperage:	6.1 A

*Increase the table value of one of the motors by 125%.

It is conceivable that the welder, all metal working lights, all service area lights, and the exhaust fan could operate at the same time, that is, without diversity.

Equipment operating without diversity	Amperage at 240 V
Welder	50.0 A
Lights: (14 lights) (1.5 A per light) = 21.0 A at 120 V	10.5 A
Exhaust fan: 7.2 A at 120 V	3.6 A
Total load:	64.1 A

Compute the demand factor at 240 volts:

Demand load at 240 V	Amperage at 240 V
Load without diversity at 100%	61.1 A
Remaining load at 50%: (0.50) (116.5 A − 64.1 A) = 26.2 A	26.2 A
Total load:	90.3 A

CHAPTER 8. EXAMPLE BUILDINGS

A 100-amp main breaker is required; the 150-amp main breaker used for this example allows for future expansion, is more cost effective to install now, and assures that the demand load is not more than about 85 percent of the main breaker. See Figure 8-3 for an SEP wiring schematic.

EXAMPLE
8-2. Livestock Building

Design the SEP for a four room farrowing building. The building is 52 × 128 feet, as illustrated in Figures 8-4 and 8-5. Each room is 32 × 46 feet, and a 6-foot wide hallway is located at the end of the rooms. Because the rooms are washed after each group of sows and pigs is removed, the system must be designed for damp conditions. Use illumination values listed in Table 1-8 as design assumptions. The SEP will be located in a utility room located next to Room #1.

SOLUTION

According to Table 1-8, the FARROWING AREA of the building should be illuminated at 10 foot-candles. Using a fixture that uses double, T8 32-watt bulb fixtures (35-watt power usage), the number of fixtures needed is:

$$\frac{(32\text{ft})(46\text{ft})\left(0.28\,\text{W}/\text{sqft}\right)}{(32+35)\,\text{W}/\text{fixture}} = 5.89 \text{ fixtures}$$

Use six, double T8 32-watt fluorescent light fixtures. Provide two fixtures over each walkway located between rows of crates.

Provide one DCO at the entry door of each farrowing room. Provide one overhead duplex receptacle SPO per two stalls for two, 250-watt heat lamps each. Wire four SPOs per individual branch circuit. Check the conductor and circuit breaker sizes:

$$\frac{\left(2\,\text{lamps}/\text{outlet}\right)\left(250\,\text{W}/\text{lamp}\right)(4\text{ outlets})}{120\text{ V}} = 16.7 \text{ amps}$$

Because these lamps will be on continuously for periods longer than 15 minutes, increase the amperage capacity by 25%:

$$(1.25)(16.7\text{ A}) = 20.8 \text{ amps, or about 20 amps}$$

The SPOs in Rooms #1 and #2 are within 60 feet of the SEP. According to Table 2-6, an AWG 8-CU conductor is needed to minimize voltage drop. Run the 8-CU from the 20 amp circuit breaker to a junction box ahead of the switches for that room, then use AWG 12-CU to the SPOs in that

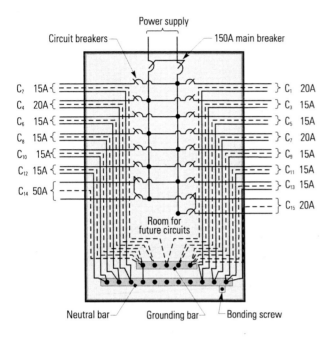

KEY:

- C_1 — 10 DCO's machine storage
- C_2 — 6 100-watt HPS lights, machine storage
- C_3 — 7 lights, T8 for metal work and metal storage
- C_4 — 6 DCO's metal work and metal storage
- C_5 — 6 lights, exterior, outside office and outlets for trouble lights
- C_6 — 6 lights T8 office, storage and exterior
- C_7 — 7 DCO's office and storage
- C_8 — 6 lights T8 wood working area
- C_9 — 6 100-watt HPS or MH lights in service area
- C_{10} — Drill 1/3 hp
- C_{11} — Exhaust fan 1/3 hp
- C_{12} — Grinder 1/3 hp
- C_{13} — Hacksaw 1/3 hp
- C_{14} — Welder 240V, 50A
- C_{15} — Office heater 240V, 5000Btu

Figure 8-3. SEP wiring schematic for example machine storage.

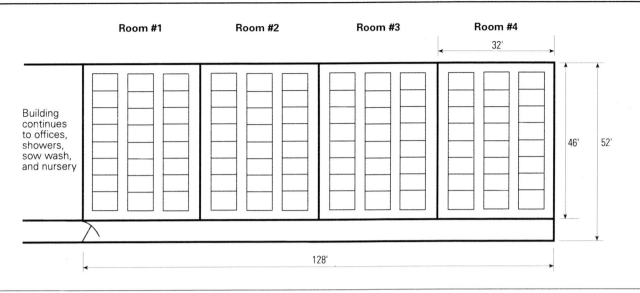

Figure 8-4. Example swine farrowing building layout.

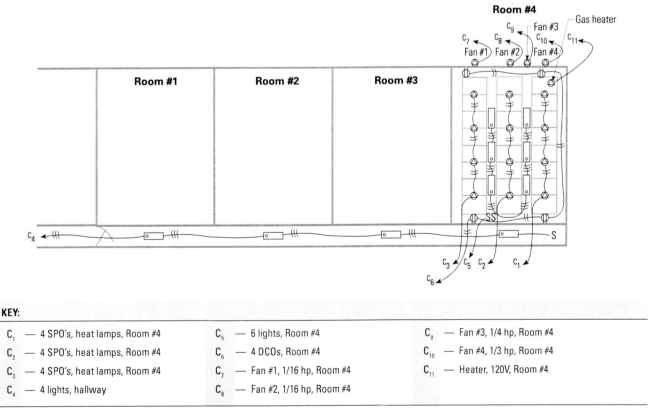

KEY:

- C_1 — 4 SPO's, heat lamps, Room #4
- C_2 — 4 SPO's, heat lamps, Room #4
- C_3 — 4 SPO's, heat lamps, Room #4
- C_4 — 4 lights, hallway
- C_5 — 6 lights, Room #4
- C_6 — 4 DCOs, Room #4
- C_7 — Fan #1, 1/16 hp, Room #4
- C_8 — Fan #2, 1/16 hp, Room #4
- C_9 — Fan #3, 1/4 hp, Room #4
- C_{10} — Fan #4, 1/3 hp, Room #4
- C_{11} — Heater, 120V, Room #4

Figure 8-5. Electrical plan for example farrowing building.

room. As listed in Table 3-1, the maximum circuit breaker capacity for AWG 12-CU conductor in UF cable is 20 amps.

The SPOs in Room #3 are within 100 feet of the SEP and within 125 feet for Room #4. An AWG-6-CU conductor is needed to minimize voltage drop. Run the 6-CU from the 20 amp circuit breaker to a junction box ahead of the switches for each room, then use AWG 12-CU to the SPOs in that room.

Place each fan and heater on a separate circuit. All the fan motors are 240 volt, and the largest is 1/3 hp with a full-load amperage of 3.6 amps, as identified by the values in Table 2-3. Because motors are sustained loads, increase the full-load current by 25% to determine conductor size:

$$(1.25)(3.6 \text{ A}) = 4.5 \text{ amps}$$

No motor in any room is more than 125 feet from the SEP, so use AWG 12-CU conductor in UF cable, as prescribed by Table 2-6. The motors are relatively small, so use combination fused switches with time-delay fuses as the disconnecting means, overload, and short-circuit protection. The maximum size for the time-delay fuses is limited to 175% of the motor's full-load current for short-circuit protection. For motor overload protection, this should not be more than 125% of motor full-load current. The circuit breaker can provide the short-circuit protection, but is not recommended.

The next step is to design the SERVICE ENTRANCE PANEL. A proposed SEP has space for 42 circuits. Each room has ten circuits that need breakers. A 120-volt circuit needs only one circuit space, but a 240-volt circuit needs two circuit spaces. Each room has four, 240-volt fans. Accounting for the extra space needed for 240-volt circuits, each room needs at least 14 circuit spaces. In addition, the hallway lights need a circuit space. A total of 57 circuit spaces are needed for the four rooms and hallway, which is more than the capacity for a single SEP. Two SEPs will be needed. One of the panels will house the breaker for the hallway lights. Each SEP will be designed for two rooms.

Motor	Size (hp)	Voltage (V)	Full-load current (A) (Table 2-2)	Maximum fuse size (A)
Fan #1	1/6	240	2.2	3.9
Fan #2	1/6	240	2.2	3.9
Fan #3	1/4	240	2.9	5.1
Fan #4	1/3	240	3.6	6.3
Heater fan	1/4	120	5.8	10.2

Electrical equipment	Amperage at 240 V
Outlets: (4 DCO / room) (2 rooms) (1.5 A per DCO) = 12.0 A at 120 V	6.0 A
Lights: (6 fixtures / room) (2 rooms) (1.5 A per fixture) + (4 hallway fixtures*) (1.5 A per fixture) = 24.0 A at 120 V	12.0 A
Heat lamps: 12 SPO per room 120 V with two 250 W heat lamps each: $\frac{(12 \text{ SPO}/\text{room})(2 \text{ rooms})(2 \text{ lamps}/\text{SPO})(250 \text{ W}/\text{lamp})}{(120 \text{ V})} = 100.0 \text{ amps}$	50.0 A
Motors: (Table 2-3), 1/6 hp at 240 V: (2 motors / room) (2 rooms) (2.2 A per motor) = 8.8 A	8.8 A
Motors: (Table 2-3), 1/4 hp at 240 V: (1 motor / room) (2 rooms) (2.9 A per motor) = 5.8 A	5.8 A
Motors: (Table 2-3), 1/3 hp at 240 V: (1.25) (1 motor** / room) (2 rooms) (3.6 A per motor) = 9.0 A	9.0 A
Motors: (Table 2-3), 1/4 hp at 120 V: (1 motor / room) (2 rooms) (5.8 A per motor) = 11.6 A	5.8 A
Total amperage:	97.4 A

* Needed for only one SEP.
** Increase the table value of the largest motor by 125%.

All the general branch circuits are arranged so no more than 10 lights, SPOs, or DCOs are connected to one circuit. General-purpose circuits require AWG 12-CU conductor and 20-amp circuit breakers. This is a **damp** building, so all fixtures and conductors in the animal area are dustproof and watertight.

Electrical equipment	Amperage at 240 V
Outlets: (4 DCO / room) (2 rooms) (1.5 A per DCO) = 12.0 A at 120 V	6.0 A
Lights: (6 fixtures / room) (2 rooms) (1.5 A per fixture) + (4 hallway fixtures*) (1.5 A per fixture) = 24.0 A at 120 V	12.0 A
Heat lamps: 12 SPO per room 120 V with two 250 W heat lamps each: $\dfrac{(12\,\text{SPO}/\text{room})(2\,\text{rooms})(2\,\text{lamps}/\text{SPO})(250\,\text{W}/\text{lamp})}{(120\,\text{V})} = 100.0$ amps	50.0 A
Motors: (Table 2-3), 1/6 hp at 240 V): (2 motors / room) (2 rooms) (2.2 A per motor) = 8.8A	8.8 A
Total load without diversity:	76.8 A

* Needed for only one SEP.

Compute the demand factor at 240 volts:

Demand load at 240 V	Amperage at 240 V
Load without diversity at 100%	76.8 A
Remaining load at 50% (0.50) (97.4 A – 76.8 A) = 10.3 A	10.3 A
Total load:	87.1 A

Use at least two, 150-amp main breaker panels for the four rooms and hallway so the demand load is not more than about 85 % of the breakers. A 150-amp main breaker is used in this example because it will allow for future expansion, is more cost effective to install now, and assures that the demand load is not more than about 85 % of the main breaker. See Figure 8-6 for an SEP wiring schematic.

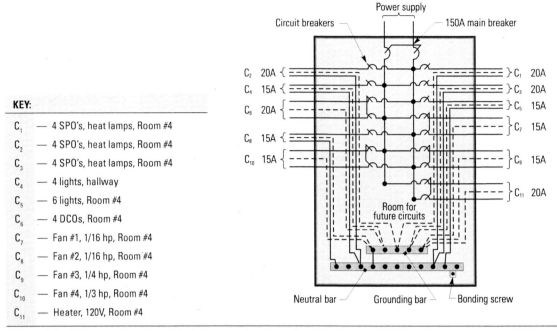

KEY:

- C_1 — 4 SPO's, heat lamps, Room #4
- C_2 — 4 SPO's, heat lamps, Room #4
- C_3 — 4 SPO's, heat lamps, Room #4
- C_4 — 4 lights, hallway
- C_5 — 6 lights, Room #4
- C_6 — 4 DCOs, Room #4
- C_7 — Fan #1, 1/16 hp, Room #4
- C_8 — Fan #2, 1/16 hp, Room #4
- C_9 — Fan #3, 1/4 hp, Room #4
- C_{10} — Fan #4, 1/3 hp, Room #4
- C_{11} — Heater, 120V, Room #4

Figure 8-6. SEP wiring schematic for example swine building.
Note: For clarity, panel shows breakers from only one room and the hallway. Another set of ten breakers, including four 240-volt breakers, will be needed.

GLOSSARY

A: Ampere(s)

ac: Alternating current

Ampacity: Allowable current carrying capacity of wire

AWG: American Wire Gage

Cable: Two or more conductors in the same protective sheath

CU: Copper

DCO: Duplex convenience outlet

GFCI: Ground-fault circuit-interrupter

hp: Horsepower

Hz: Hertz

IEEE: Institute of Electrical and Electronic Engineers

kW: Kilowatt

NEC: National Electrical Code

NEMA: National Electrical Manufacturers Association

NESC: National Electrical Safety Code

NFPA: National Fire Protection Association

o.c.: On-center

PTO: Power-take-off

PVC: Polyvinyl chloride

SEP: Service entrance panel

SPO: Special purpose outlet

UL: Underwriters' Laboratories

USE: Underground service entrance

V: Volts(s)

W: Watt(s)

Wire: A single conductor

ADDITIONAL REFERENCES

American Society of Agricultural and Biological Engineers, 2950 Niles Road, St. Joseph, MI 49085 (www.asabe.org):
- *Equipotential Plane in Livestock Containment Areas*, EP473.2
- *Lighting Systems for Agricultural Facilities*, EP344.3
- *Safety for Electrically Heated Livestock Waterers*, EP342.3
- *Specifications for Alarm Systems Utilized in Agricultural Structures*, S417.1
- *Specifications for Lightning Protection*, EP381.1
- *Installation and Maintenance of Farm Standby Electric Power*, EP364.3

Electrical Generating Systems Association (EGSA), 1650 S Dixie Hwy, Ste 400, Boca Raton, FL 33432, (www.egsa.org).

Lightning Protection Institute, P.O. Box 6336 St. Joseph, MO 64501, (www.lightning.org):
- *Standard of Practice for the Design-Installation-Inspection of Lightning Protection Systems*

National Fire Protection Association (NFPA), 1 Batterymarch Park, Quincy, Massachusetts, 02169-7471 (www.nfpa.org):
- *2011 National Electrical Code*, NFPA 70

Standard for the Installation of Lightning Protection Systems, NFPA 780, 2004 Edition

Rural Electricity Resource Council, P.O. Box 309, 2333 Rombach Ave., Wilmington, OH 45177 (www.nfec.org):
- *Agricultural Wiring Handbook*
- *Electrical Wiring for Livestock and Poultry Structures*
- *Planning Electrical Needs for Crop Drying and Storage*

Underwriters Laboratories Inc., 333 Pfingsten Road, Northbrook, IL 60062-2096, (www.ul.com):
- *General Information for Electrical Equipment*, 2008

INDEX

A
Alarm systems, 53-55
Amperage, 32-34, 37
AC snap switch, 29

B
Bonding strip (jumper), 41
Boxes, 4-8
Branch circuits, 19-34
    120-V circuits, 19-20
    240-V circuits, 19-20
    Conductors, 32-34
    Circuit types, 20-21
    Sizing, 32-34
Building groups, 3-4, 11

C
Circuit, 19
    Breakers, 35-36
    Neutrals, 19
Codes, 1-2
Common grounding 74
Conductance, bodies of, 73
Conductors, 3, 6-9
Counterpoise ground, 76
Controller, motor circuits, 23-26, 29-31, 82

D
Demand factor, 39
Disconnecting means, 23-26, 28-29, 82
Duplex convenience outlet (DCO), 15, 18, 20-21

E
Equipment grounding, 19
Equipotential planes, 65, 67-70

G
Generator, 47-52
Grounding, 19, 40-43
Ground fault circuit interrupters, 35-36
Grounding electrode conductor, 41-42
Grounding electrode system, 42
Ground-faults, 59

I
Illumination levels, 16-17
Inductance, bodies of, 73-74
Isolation transformers, 64-66

K
Knife switch, 29

L
Light levels, 16-17
Lightning protection, 71-77
    Air terminals, 72
    Arresters, 74-75
    Fences, 77
    Ground connections, 75-76
    Main conductors, 72-73
    Metal-clad and steel-framed buildings, 76
    Secondary conductors, 73-74
    Trees, 77
Lights, 12-18
    Characteristics, 12-14
    Compact fluorescent (CFL), 14
    Fluorescent, 12-14
    High intensity discharge (HID), 14-15
    Illumination levels, 16-17
    Incandescent, 12

M
Main breaker, 37-39
Materials, 2-11
    Boxes, 4-6
    Damp buildings, 4-10, 22
    Dry buildings, 3
    Dusty buildings, 11
    Plastic conduit, 8-10
    Surface mount cable, 4
Motor circuits, 22-32
    Controller, 23-26, 29-31
    Disconnecting means, 23-26, 28-29
    Overload protection, 23-26
    Short-circuit protection, 27-28

N

National Electrical Code (NEC), 1
National Electrical Safety Code (NESC), 1-2

O

Outlets, 15, 18, 20-21
Overload protection, motors, 31-32, 82

P

Planning, 11
Power to building, 44-45

S

Safety, 1-2
Service entrance, 36-42, 80-88
    Conductors, 43-44
    Ground, 42-44
    Ground fault circuit interrupters, 35-36
    Grounding electrode conductor, 41-42
    Grounding electrode system, 42
    Installing circuit breakers, 35
    Main breaker, 37-40
    Panel, 36-41, 80-88
    Sizing circuit breakers, 35
    Wiring diagram, 40

Short-circuit protection, motors, 27-28, 82
Special purpose outlets (SPO), 15, 18, 20-21
Standby power, 47-52
Stray voltage, 57-70
    Causes, 58
    Eliminating or reducing, 64-65
    Equipotential planes, 65, 68-70
    Isolating voltage, 64
    Off-farm sources, 62-63
    On-farm sources, 59-62
Sub panels, 37
System grounding, 19, 40-44

U

Underground wiring, 45
Underwriters Laboratories Inc. (UL), 2

V

Voltage, 33-34

W

Wattage, 33-34